The Virtues of Limits

The Virtues of Limits

DAVID MCPHERSON

OXFORD

UNIVERSITY PRESS

Great Clarendon Street, Oxford, OX2 6DP,
United Kingdom

Oxford University Press is a department of the University of Oxford.
It furthers the University's objective of excellence in research, scholarship,
and education by publishing worldwide. Oxford is a registered trade mark of
Oxford University Press in the UK and in certain other countries

Published in the United States of America by Oxford University Press
198 Madison Avenue, New York, NY 10016, United States of America

British Library Cataloguing in Publication Data
Data available

Library of Congress Cataloging in Publication Data
Data available

ISBN 978–0–19–284853–6 (Hbk.)
ISBN 978–0–19–896965–5 (Pbk.)

DOI: 10.1093/oso/9780192848536.001.0001

Printed and bound by CPI Group (UK) Ltd, Croydon, CR0 4YY

Links to third party websites are provided by Oxford in good faith and
for information only. Oxford disclaims any responsibility for the materials
contained in any third party website referenced in this work.

The manufacturer's authorised representative in the EU for product safety is Oxford University Press España S.A.
of El Parque Empresarial San Fernando de Henares, Avenida de Castilla, 2 – 28830 Madrid (www.oup.es/en or
product.safety@oup.com). OUP España S.A. also acts as importer into Spain of products made by the manufacturer.

For Kirstin

Contents

Acknowledgments   ix

Introduction   1

1. Existential Limits   5
   The Promethean Ideal   5
   Against the Promethean Ideal   15
   Humility and Reverence   28
   Contentment and Gratitude   34
   Loyalty to the Given   40

2. Moral Limits   46
   Character Formation: Beginning with Restraint,
      Cultivating Reverence and Moderation   46
   Absolute Prohibitions and the Virtue of Reverence   62
   Duties of Assistance: Neighborliness and Loyalty
      as Limiting Virtues   71

3. Political Limits   84
   The Bonds and Bounds of Political Community   84
   Sufficientarian Justice   98
   Against Utopianism: The Politics of Imperfection   109
   Moderation and the Limits of Government   119

4. Economic Limits   126
   The Vice of Greed and the Virtue of Contentment
      in Economic Life   126
   Home Economics   143
   A Sabbath-Orientation   155

Works Cited   163
Index   173

Acknowledgments

I would like to express my gratitude to Creighton University for a Dr. George F. Haddix President's Faculty Research Fund Grant that supported my work on this book. I also thank the Creighton College of Arts and Sciences for supporting my research, and my colleagues Ross Romero, Patrick Murray, Jeanne Schuler, and Anne Ozar for many good discussions. Additionally, I thank my students in my courses who have helped me to think through many of the issues discussed here. Special thanks are due to Jack Eastman, with whom I did an independent study on virtue ethics and economics; our readings and discussions helped me to think through the issues discussed in Chapter 4.

I am blessed with wonderful friends in philosophy, and many of them were very generous with their time in reading and commenting on all or parts of the manuscript. For their helpful comments, I offer my heartfelt thanks to Kirstin McPherson, John Cottingham, Tom Angier, Anthony O'Hear, Edward Skidelsky, Simon May, Ryan Hanley, Richard Kim, Cora Diamond, Patrick Murray, and Tristan Rogers. I am also grateful for helpful conversations with Fiona Ellis, Brad Cokelet, Jennifer Frey, Brandon Warmke, Jason Baehr, Greg Beabout, Sam Fleischacker, Mathew Lu, Chris Toner, Patrick Toner, and Roger Scruton. Sadly, Roger passed away in January 2020. I would very much have liked for him to have seen the finished book. I hope he would have liked it, since his work has been an important influence, and he had been a source of encouragement. May he rest in peace.

I want to thank audiences at the University of Cape Town in Cape Town, South Africa, and at the Research Institute for Politics and Government at the National University of Public Service in Budapest, Hungary, for helpful feedback on material related to this book. I also thank the other participants at a Liberty Fund conference on "Liberty and the Sabbath" in Santa Fe, New Mexico. Thanks to Daniel Ritchie for organizing the conference. This was a helpful forum for working out my thoughts on the Sabbath, which are discussed at the end of the book.

I am grateful for the permission to make use of some material from my essay "Existential Conservatism" (*Philosophy* 94:3, 2019) in Chapters 1 and 3. I am also grateful for the permission to make use of some material from my essay "Manners and the Moral Life" (in *The Theory and Practice of Virtue Education*, ed. Tom Harrison and David Walker [New York: Routledge, 2018]) in Chapter 2 as well as from my essay "Nietzsche, Cosmodicy, and the Saintly Ideal" (*Philosophy* 91:1, 2016) in Chapter 1.

I would like to offer special thanks to my editor Peter Momtchiloff at Oxford University Press for his support of this project. I also thank everyone who has helped to bring this book to press, especially Céline Louasli and Thomas Deva.

Lastly, I want to express my deep gratitude to my family. Specifically, I thank my parents for all their love and support over the years. Above all, I thank my wife Kirstin and our children, Clare, John, Peter, and Andrew, for the incalculable blessing they have been in my life. I write at home amidst family life, and my work is shaped by our life together as a family. For instance, my reflections on contentment in Chapter 4 were influenced by reading and discussing Patricia Polacco's *Luba and the Wren*—a Ukrainian fairy tale akin to the Brothers Grimm's *The Fisherman and His Wife*—with my daughter Clare (then 7 years old), who is already quite the philosopher. Almost everything I write I have discussed with Kirstin as part of working out together our shared world-view, and I have greatly benefited from all our conversations. I have dedicated this book to her as a small token of my love and appreciation for her and the great blessing she has been to me. She is my dearest friend and companion in all things.

Introduction

Human beings seek to transcend limits. This is part of our potential greatness, since it is how we can realize what is best in our humanity. However, the limit-transcending feature of human life is also part of our potential downfall, as it can lead to dehumanization and failure to attain important human goods and to prevent human evils. In this book I explore the place of limits within a well-lived human life and develop and defend an original account of what I call "limiting virtues," which are concerned with recognizing proper limits in human life.[1] The limiting virtues that are my focus are humility, reverence, moderation, contentment, neighborliness, and loyalty. These virtues have been underexplored in discussions about virtue ethics, and when they have been explored, it has not been with regard to working out a position on the general issue of the place of limits within a well-lived human life. The account of the limiting virtues provided here, however, is intended as a counter to other prominent approaches to ethics, namely, autonomy-centered approaches and consequentialist (or maximizing) approaches. I develop and defend my account of the limiting virtues in relation to four kinds of limits: (1) existential limits; (2) moral limits; (3) political limits; and (4) economic limits. The four chapters of this book correspond to each of these types of limits.

On my view the virtues are modes of proper responsiveness to that which is of intrinsic value (or goodness) and which makes normative demands upon us, and in being properly responsive the virtues constitute for us the good life, that is, our human fulfillment understood as a

---

[1] The title of this book is intended to have a double meaning. In one sense, "the virtues of limits" is another way of speaking about "the limiting virtues." However, in another sense, "the virtues of limits" suggests more broadly that there are benefits of recognizing a proper place for limits in human life.

The Virtues of Limits. David McPherson. Oxford University Press. © David McPherson 2022.
DOI: 10.1093/oso/9780192848536.003.0001

normatively higher, nobler, more meaningful form of life.[2] In a general sense then all of the virtues—e.g., courage or generosity—can be understood as having a limiting function in so far as in being properly responsive to intrinsic values—e.g., human dignity or the nobility of virtue itself in realizing what is admirable in our humanity—we recognize constraints on our desires and choices. However, the limiting virtues that I discuss recognize limits in more specific ways in relation to the four kinds of limits that I have mentioned.

Humility can be regarded as the master limiting virtue: it ensures that we recognize and live out our proper place in the scheme of things. As a limiting virtue, it is especially concerned with reining in what—in Chapter 1—I describe as the Promethean tendency to "play God" in seeking mastery over the given world, which is a prominent tendency in the modern world that is seen especially in a certain scientific-technological mindset. The virtue of humility recognizes that some things must be accepted and appreciated *as given*, and not subject to human control or manipulation. It properly acknowledges our dependency on others and on the natural world, as well as on values (or goods) not of our own making for living well and meaningfully as human beings. The virtue of humility also properly acknowledges our natural, personal, and moral limitations.

The limiting virtue of reverence is concerned with being properly responsive, through reverential attitudes and behavior, to that which is reverence-worthy (e.g., human life) and which places strong constraints on our will. As I discuss in Chapter 1, the virtue of reverence is closely connected with humility because being properly responsive to that which is reverence-worthy helps to define our proper place in the scheme of things. In Chapter 2 I develop a Confucian account of the importance of reverence within character formation, and I also argue for its importance for recognizing absolute moral prohibitions.

In Chapter 2 I also put forward an Aristotelian account of the importance of moderation within character formation. Moderation is a limiting virtue because it is concerned with avoiding vicious extremes. It can

---

[2] For more on this view, see David McPherson, *Virtue and Meaning: A Neo-Aristotelian Perspective* (Cambridge: Cambridge University Press, 2020), ch. 2.

also be understood as a master virtue in so far as the virtues of character consist in realizing a proper mean between excess and deficiency in some feeling or action. However, in Chapter 2 I focus on moderation in the form of temperance and discuss how it enables us not to be enslaved to our animal appetites and makes us receptive to that which is ennobling of our humanity. In Chapter 3 I also discuss the importance of moderation in the political domain as part of a politics of imperfection and how it is especially important for helping us to deal with the problem of conflict within political community.

The limiting virtue of contentment is the virtue of knowing when enough is enough, of not wanting more than is needed for a good life. It does not deny that we ought in many ways to seek improvement, but it acknowledges—as I discuss in Chapter 1—that we need to find a way to be *at home* in the given world amidst imperfection. This requires that we cultivate a grateful or appreciative orientation toward the world. In Chapter 3 I also discuss the role of contentment in a politics of imperfection and its connection with the sufficientarian account of distributive justice that I defend, where what is important is that people have enough to live well. In Chapter 4 I discuss the importance of contentment for counteracting the vice of greed, and I connect it with a vision of economic life that contributes to being at home in the world.

The limiting virtue of neighborliness is a form of human solidarity that recognizes the moral significance of proximity. It stands opposed to impartialist moral theories, such as utilitarianism and Kantianism, which do not recognize the moral significance of proximity. While it has been overlooked or disregarded by such moral theories, the virtue of neighborliness has had a prominent place in Western culture due to the influence of the biblical teachings regarding love of neighbor. As we see in the Parable of the Good Samaritan, our neighbor whom we are to love is not just someone who lives nearby and who is part of our community, but anyone—including strangers—we encounter face to face. In Chapter 2 I discuss how the virtue of neighborliness, with its focus on concrete rather than abstract humanity, should inform how we think about duties of assistance. In Chapter 3 I also discuss its importance for how we think about the bonds and bounds of political community.

When we love and care for those who are *there* in our lives, we will form identity-constituting bonds of attachment with some of these particular people and will come to recognize demands of loyalty to them that sustain the good of the relationship and which give grateful recognition to the good we have received from them. The virtue of loyalty is a limiting virtue that expresses proper partiality, and thus it places limits on the extent of our attachments and how far we can be expected to go in pursuing impartial concern. It involves binding attachment that is maintained through thick and thin (i.e., for better or for worse). In Chapter 2 I discuss loyalty to friends and family, and in Chapter 3 I discuss patriotism, which involves loyalty to one's country and fellow citizens. In Chapter 1 I also discuss what I call "loyalty to the given," which recognizes that the given world places demands upon us for loyalty, and we fail to be properly responsive to existing value by refusing to belong to the given world. This loyalty to the given provides the wider context in which more particular loyalties find their proper place.

What emerges over the course of the four chapters of this book is a broad and distinctive vision of the good life that recognizes the importance of limits and the limiting virtues for living well as human beings. It is my hope that the broad vision of the good life offered here is one that the reader will find compelling.

1

Existential Limits

In exploring the place of limits within a well-lived human life I will begin at the most general level with existential limits. By "existential limits" I mean limits with respect to *the given*, that is, to what exists. In this chapter I want to explore two fundamental *existential stances*, or *orientations toward the given*, which will play a key role throughout the book. One stance we can take toward the given is *the choosing-controlling stance*. All mature human beings adopt this stance to some extent in their efforts to improve their lives and the world around them through controlling, transforming, and overcoming the given. However, at the extreme, this stance can give expression to a "Promethean" project of "playing God" by seeking mastery over the given. The other basic existential stance is at odds with this Promethean project: I refer to it as *the accepting-appreciating stance*. By accepting and appreciating the given, it imposes limits on the choosing-controlling stance. While acknowledging an important role for a choosing-controlling stance in human life, I will argue that the accepting-appreciating stance toward the given should be regarded as primary in several ways, and I will discuss specific limiting virtues—namely, humility, reverence, contentment, and loyalty—that give proper expression to this stance. But to make this case we first need to consider the rival view that seeks to make the choosing-controlling stance dominant by aspiring to a Promethean ideal of mastery.

The Promethean Ideal

In the ancient Greek myth, Prometheus—who is a Titan, a non-Olympian god—steals fire from the Olympian gods and gives it to

The Virtues of Limits. David McPherson. Oxford University Press. © David McPherson 2022.
DOI: 10.1093/oso/9780192848536.003.0002

humanity.[1] The possession of fire here can be understood as a godlike power to transform the natural world, and thus it can also be understood as a symbol of civilization in so far as this is seen as being advanced through science and technology.[2] The myth of Prometheus is thus often interpreted as offering us a godlike ideal of human mastery over nature, and so the Promethean ideal can be understood as an ideal of "playing God."[3]

The tendency to play God is as old as humanity. Indeed, there are other notable instances of it in the ancient world, such as in the biblical story of the Tower of Babel (Genesis 11:1–9). But this tendency becomes especially prominent in the modern world. The story of modernity can, in fact, be told in terms of the Promethean ideal of mastery, which finds expression in a prominent scientific-technological mindset, from Francis Bacon and René Descartes onward.[4] According to this mindset, we

---

[1] The first account of the myth of Prometheus that we have is in Hesiod's *Theogony*, and it is discussed further in his *Works and Days*. It is also recounted in Aeschylus' *Prometheus Bound*, and indeed throughout the ages.

[2] William Hansen writes of the myth of Prometheus:

> [Without] fire men could not convert the flesh into food suitable for human beings. Indeed, without fire men were deprived even of the ability to manufacture utensils and tools, for they could neither fire clay pots nor smelt metals, nor could they employ fire for warmth and light. In short, humans would live no better than animals.... The Titan [Prometheus] made the divine element of fire permanently available to men by placing it in trees, for once fire was stored in wood, men could extract it by friction whenever it suited them. Human beings now possessed fire for cooking, heat, light, and technology. Their life would be above that of the animals, but of course below that of the gods.
>
> (*Classical Mythology: A Guide to the Mythical World of the Greeks and Romans*, 2nd edn [Oxford: Oxford University Press, 2020], 62)

[3] David E. Cooper writes:

> The figure of the Titan [Prometheus] who, in Aeschylus's *Prometheus Bound*, rejects "servile humility" to Zeus and refuses to "speak humbly and fear Nemesis," is an apt symbol for a philosophy which seemingly denies the answerability of human thought and purpose to any independent measure which might prove to be their nemesis.... [As] bearer of fire to human beings and, in Aeschylus's words, the giver of "all human skills and science," Prometheus is a symbol of the practical and technical dimensions of the human enterprise.
>
> (*The Mystery of Things: Humanism, Humility, and Mystery* [Oxford: Oxford University Press, 2002], 52)

[4] See William Leiss, *The Domination of Nature* (Boston, MA: Beacon, 1972); Rémi Brague, *The Kingdom of Man: Genesis and Failure of the Modern Project*, trans. Paul Seton (Notre Dame, IN: University of Notre Dame Press, 2018); Charles Taylor, *Sources of the Self: The Making of the Modern Identity* (Cambridge, MA: Harvard University Press, 1989), 84–6, chs. 8 and 13; John Cottingham, "Plato's Sun and Descartes's Stove: Contemplation and Control in Cartesian Philosophy," in *Cartesian Reflections: Essays on Descartes's Philosophy* (Oxford: Oxford University Press, 2008), 292–318.

should not be passive to nature, but rather we should seek to master it through science and technology and direct it, as far as possible, toward the fulfillment of our desires. This often goes together with a disenchanted understanding of nature where it is bleached of intrinsic values and moral purposes that would place normative constraints on our will. Besides this scientific-technological mindset, there are also strands of Prometheanism that focus on human powers of creativity, especially in the realm of art. But in each case, what we have is a primacy of the choosing-controlling stance.[5]

I want to focus here on two efforts to take up the Promethean project of playing God in making the choosing-controlling stance primary. One is that of Friedrich Nietzsche in the late nineteenth century, who is more in line with the strand that emphasizes human powers of creativity, and the other is a more recent case, namely, Ronald Dworkin's defense of genetic engineering, which is expressive of the scientific-technological version of Prometheanism.

Nietzsche is arguably the most famous modern proponent of the Promethean ideal. We see it most clearly in his "death of God" passage in *The Gay Science*.[6] Here "the madman" (who is a precursor to Nietzsche's Zarathustra in *Thus Spoke Zarathustra*) proclaims the "death of God," and indeed he says: "We have killed him" (§125). Nietzsche later explains that the "death of God" means that "belief in the Christian god has become unbelievable" (§343). And so on his view, we—i.e., we "good Europeans [or moderns]"—have "killed God" in that we have made God unbelievable, especially through modern science:

---

[5] I should note that I think the best theological view would maintain that God does not "play God" in the sense of taking a stance of mastery that recognizes no intrinsic values in the world that place constraints on the will. On this view, it is part of God's holiness that God is properly responsive to what is of intrinsic value. The Promethean ideal of playing God is more akin to the voluntarist view of God that came into prominence in the late medieval period, which maintains that good is good *because* it is willed by God, which contrasts with traditional theistic natural law ethics (affirmed by Aquinas and others), which maintains that God wills what is intrinsically good on the basis of a perfect understanding of the good. Because of its connections with nominalism, the voluntarist view in fact is often understood to have contributed to the rise of the modern scientific-technological mindset.

[6] Friedrich Nietzsche, *The Gay Science*, trans. Walter Kaufmann (New York: Vintage Books, 1974 [1882/1887]).

> Looking at nature as if it were proof of the goodness and governance of
> a god; interpreting history in honor of some divine reason, as a
> continual testimony of a moral world order and ultimate moral
> purposes; … that is *all over* now, that has man's conscience *against* it,
> that is considered indecent and dishonest by every more refined
> conscience.   (§357)

Belief in God, in fact, has not gone away, nor is it likely to do so in the
future, and for good reason, given the sort of ultimate questions to which
it responds, which themselves are not going away.[7] But Nietzsche is right
that there is a challenge here, and my concern is to consider what he
regards as the implications of the death of God.

The first response to the realization of the death of God that Nietzsche
describes is the experience of existential vertigo. The madman says:

> What were we doing when we unchained this earth from its sun?
> Whither is it moving now? Whither are we moving? Away from all
> suns? Are we not plunging continually? Backward, sideward, forward,
> in all directions? Is there still any up or down? Are we not straying as
> through an infinite nothing? Do we not feel the breath of empty space?
> Has it not become colder? Is not night continually closing in on us?
>
> (§125)

In other words, the death of God means a fundamental shift in our
conception of the given, which induces a feeling of existential disorientation along with a need to re-evaluate our values. Nietzsche writes:
"[How] much must collapse now that this faith [in God] has been
undermined because it was built upon this faith, propped up by it,
grown into it; for example, the whole of our European morality"
(§343). There is no longer "a moral world order and ultimate moral
purposes" to which we must align our lives, and so we must find another
way to orient our lives if we are not to sink into nihilism and despair.
And the madman suggests a solution to the problem of the death of God:
we must play God. As he puts it: "Is not the greatness of this deed too

---

[7] See John Cottingham, *On the Meaning of Life* (New York: Routledge, 2003), chs. 1 and 2.

great for us? Must we ourselves not become gods simply to appear worthy of it?" (§125).

And with this possibility there can be a second response to the death of God, namely, a sense of liberation, which Nietzsche thinks is on the other side of the experience of existential vertigo. In a section titled "The meaning of our cheerfulness," Nietzsche writes:

> [We] philosophers and "free spirits" feel, when we hear the news that "the old god is dead," as if a new dawn shone on us; our heart overflows with gratitude, amazement, premonitions, expectation. At long last the horizon appears free to us again, even if it should not be bright; at long last our ships may venture out again, venture out to face any danger; all the daring of the lover of knowledge is permitted again; the sea, *our* sea, lies open again; perhaps there has never yet been such an "open sea."
>
> (§343)

There is an exhilarating experience of a certain kind of *limitlessness* here: we are no longer ultimately accountable to God; nor are we constrained by any inherent moral order not of our own making. In this sense, for Nietzsche, "everything is permissible" (as Ivan Karamazov says). We are on our own in the universe and can make of our lives what we will, that is, we are free to engage in projects of *self-creation*. Our lives can thus be seen as works of art, which we "sculpt" along with the values that guide them through our human powers of creativity.[8]

But this does not mean that Nietzsche thinks there is no best way to live, since he thinks some lives are more existentially viable than others.

---

[8] This is a theme of Nietzsche's work from early on, as seen in *The Birth of Tragedy*: "[We] have our highest dignity in our significance as works of art—for it is only as an *aesthetic phenomenon* that existence and the world are eternally *justified*" (trans. Walter Kaufmann, in *Basic Writings of Nietzsche* [New York: The Modern Library, 2000 (1872)], §5; see also *The Gay Science* §107). Later, in *The Gay Science*, he writes: "*One thing is needful.*—To 'give style' to one's character—a great and rare art!" (§290). Also:

> Let us therefore *limit* ourselves to the purification of our opinions and valuations and to the *creation of our own new tables of what is good*, and let us stop brooding about the "moral value of our actions"!…We…*want to become those we are*— human beings who are new, unique, incomparable, who give themselves laws, who create themselves.   (§335).

Again, we still must be able to ward off the threat of nihilism and despair, and as we have seen, Nietzsche's recommended solution to this problem is to play God. This brings us to Zarathustra's teachings on the "over-man" (*Übermensch*) in *Thus Spoke Zarathustra*.[9] In madman-like fashion, Zarathustra says:

> *I teach you the overman.* Man is something that shall be overcome. What have you done to overcome him? All beings so far have created something beyond themselves; and do you want to be the ebb of this great flood and even go back to the beasts rather than overcome man? What is the ape to man? A laughingstock or a painful embarrassment. And man shall be just that for the overman: a laughingstock or a painful embarrassment…. The overman is the meaning of the earth. Let your will say: the overman *shall be* the meaning of the earth!…Man is a rope, tied between beast and overman—a rope over an abyss…. What is great in man is that he is a bridge and not an end.   ("Prologue" §§3, 4)

Here we seem to have a transhumanist ideal that is said to give meaning to our lives (and indeed to "the earth") in so far as we act as a stepping stone (or "bridge" or "rope") to the achievement of this ideal of the "overman," that is, the godlike individual who transcends the mediocre or "human, all too human" level of achievement by achieving something truly great. Nietzsche is before the time of the sort of genetic engineering that we will see Dworkin endorse, and so for him the production of great individuals is primarily a matter of creative cultural achievement.

Nietzsche puts forward what can be described as an ethic of power; in his book *The Antichrist*—which opposes the Christian ethic of love—he writes: "What is good?—Everything that heightens the feeling of power in man, the will to power, power itself. What is bad?—Everything that is born of weakness. What is happiness?—The feeling that power is *growing*, that resistance is overcome."[10] The ideal of the overman can thus be seen as an expression of this ethic of power: we have an ideal here of

---

[9] Friedrich Nietzsche, *Thus Spoke Zarathustra*, trans. Walter Kaufmann (New York: Modern Library, 1995 [1883–5]).

[10] Friedrich Nietzsche, *The Antichrist*, trans. Walter Kaufmann, in *The Portable Nietzsche* (New York: Penguin, 1954 [1888]), §2.

"self-overcoming" through continual growth in power, whereby one moves toward ever more enhanced forms of achievement in overcoming limitations or "resistances."[11] For Nietzsche, this ideal of the overman is central to his attempt to overcome what he sees as the nihilism of Arthur Schopenhauer, who, as is well-known, was a key influence on his philosophy. Nietzsche regards Schopenhauer as saying no to life through his stance of resignation because of his belief in the pervasive and ineliminable reality of suffering in human life, due especially to the insatiability of human desire. Contra Schopenhauer, Nietzsche maintains that the pervasiveness and ineliminability of suffering in human life is no reason to reject the "will to life"; rather, suffering should be affirmed as integral to the process of overcoming resistance and achieving greatness in the expression of the will to power; for example, in artistic, athletic, intellectual, or political achievement. Nietzsche also takes issue with Schopenhauer's morality of compassion and the similar sort of morality that is embraced in modern humanitarianism that regards suffering as something simply to be eradicated. He regards this morality of compassion as the "practice of nihilism" as it expresses "hostility to life," since life is necessarily bound up with suffering.[12] He also regards such humanitarianism as part of a broader leveling tendency in modern democratic societies that finds expression in other positions such as egalitarianism, utilitarianism, and hedonism; each of these stands opposed to Nietzsche's concern for achieving greatness in the expression of the will to power.[13]

In *Thus Spoke Zarathustra*, Nietzsche encapsulates this modern leveling tendency in the figure of the "last man," who is antipodal to his ideal of the overman:

> Alas, the time is coming when man will no longer shoot the arrow of his longing beyond man, and the string of his bow will have forgotten how to whir! . . . Behold, I show you the *last man*. "What is love? What

---

[11] For more on this, see David McPherson, "Nietzsche, Cosmodicy, and the Saintly Ideal," *Philosophy* 91:1 (2016): 48–51. See also Bernard Reginster, *The Affirmation of Life: Nietzsche on Overcoming Nihilism* (Cambridge, MA: Harvard University Press, 2006), ch. 3. Reginster's book has influenced my reading of Nietzsche here.

[12] Nietzsche, *The Antichrist*, §7.   [13] See, e.g., Nietzsche, *The Gay Science*, §377.

is creation? What is longing? ..." thus asks the last man, and he blinks. The earth has become small, and on it hops the last man, who makes everything small .... One still loves one's neighbors and rubs against him, for one needs warmth .... Everybody wants the same, everybody is the same ... One has one's little pleasure for the day and one's little pleasure for the night: but one has regard for health. "We have invented happiness," say the last men, and they blink.   ("Prologue" §5)

Nietzsche describes the happiness of the last man as "wretched contentment," and he says one's "happiness ought to justify existence itself" ("Prologue" §3), but the last man's happiness clearly does not, whereas he thinks the overman's happiness—which involves the "feeling that power is *growing*, that resistance is overcome"—does. As we saw Zarathustra proclaim: "The overman is the meaning of the earth." In other words, existence as a whole is justified in that it has given rise over the course of evolution to individuals who achieve greatness in the expression of the will to power, which amounts to making the choosing-controlling stance primary.

Even though Nietzsche rejects humanitarianism, it is possible to make the choosing-controlling stance toward the given primary in service of the humanitarian goal of improving quality of life—as we see with the modern scientific-technological mindset—and here there is anything but contentment in the restless attempt to improve quality of life. Indeed, as we see in discussions around genetic engineering today, the effort to improve quality of life has led to a transhumanitarian goal, which seeks to transcend the "human, all too human."

Let us turn then to consider Ronald Dworkin's defense of playing God in his discussion of genetic engineering of children, where this involves choosing which kind of human beings there will be by "altering a zygote's genetic composition to produce a battery of desired physical, mental, and emotional propensities."[14] Dworkin notes that the prospect of this has

---

[14] Ronald Dworkin, "Playing God: Genes, Clones, and Luck," in *Sovereign Virtue: The Theory and Practice of Equality* (Cambridge, MA: Harvard University Press, 2000), 438. Other defenses of "playing God" with genetic engineering can be found in Julian Savulescu and Nick Bostrom (eds.), *Human Enhancement* (Oxford: Oxford University Press, 2009) and Steve Clarke, Julian Savulescu, C. A. J. Coady, Alberto Giubilini, and Sagar Sanyal (eds.), *The Ethics of Human*

aroused great fear and revulsion, and he explores the possible reasons for this, such as fear of possible bad consequences (e.g., increased miscarriages or deformities), concerns about justice (e.g., that genetic engineering will exacerbate the divide between rich and poor, as the former will obtain even further advantages over the latter), and aesthetic worries about a loss of human diversity. Dworkin believes that these concerns can be addressed or are overblown, and that, separately or taken together, they do not explain the strength of the reaction against genetic engineering. He thinks that we can come to understand better this reaction by getting clear on the commonly voiced objection that genetic engineering involves a wrongful attempt to "play God," which is thought to be "wrong in itself, quite apart from any bad consequences it will or may have for any identifiable human being" (443). Dworkin says that it is often unclear what the injunction "Don't play God" is supposed to mean. It cannot mean that we should not resist natural disasters and diseases (e.g., by building a levee or using medicine) and improve the hand we have been dealt by nature (e.g., through strenuous exercise), since human beings have always done so.

To make sense of the playing God objection, Dworkin says, we need to see that the overall structure of our ethical experience "depends, crucially, on a fundamental distinction between what we are responsible for doing or deciding, individually or collectively, and what is given to us, as background against which we act or decide, but which we are powerless to change" (443). He goes on to write:

[The] crucial boundary between chance and choice is the spine of our ethics..., and any serious shift is seriously dislocating.... We dread the prospect of people designing other people because that possibility

---

Enhancement: Understanding the Debate (Oxford: Oxford University Press, 2016). There are other possible forms of genetic engineering that do not involve making "designer babies," such as therapeutic uses of gene editing through CRISPR-Cas9 technology on already born human beings (there is also genetic engineering of non-human life). In my discussion of genetic engineering here I am focusing on the "designer baby" issue, since it seems to be a clear case of Prometheanism. For a helpful volume discussing the ethical issues surrounding new gene editing technology (including with application to the genetic engineering of children), see Erik Parens and Josephine Johnston (eds.), *Human Flourishing in an Age of Gene Editing* (Oxford: Oxford University Press, 2019).

in itself shifts ... the chance/choice boundary that structures our values as a whole, and such a shift threatens, not to offend any of our present values ..., but, on the contrary, to make a great part of these suddenly obsolete.... The terror many of us feel at the thought of genetic engineering is not a fear of what is wrong; it is rather a fear of losing our grip on what is wrong.   (444, 446)

In his attempt to make a choosing-controlling stance toward the given dominant over an accepting-appreciating stance, we can see Dworkin, like Nietzsche, having to confront an experience of existential disorientation, which also involves a lurking threat of nihilism, that is, that nothing really matters. But he sees this as a challenge to confront with courage rather than a reason for turning back; it is an "open sea" to venture out upon, as Nietzsche put it. Dworkin writes: "Playing God is indeed playing with fire. But that is what we mortals have done since Prometheus, the patron saint of dangerous discovery. We play with fire and take the consequences, because the alternative is cowardice in the face of the unknown" (446).

However, unlike Nietzsche, Dworkin does not believe that this necessarily calls for a complete revaluation or turning upside down of currently held values, at least not those of his own liberal egalitarian outlook, which he describes as "ethical individualism." It has two key principles. The first principle "holds that it is objectively important that any human life, once begun, succeed rather than fail" (448). In other words, we have here a principle of beneficence concerned with promoting good quality of life for all human beings. The second principle insists that "one person—the person whose life it is—has a special responsibility for each life, and that in virtue of that special responsibility he or she has a right to make the fundamental decisions that define, for him, what a successful life would be" (449). In other words, we have here the standard liberal principle of autonomy, where people should be free to do what they want so long as they do not infringe on the ability of others to do likewise. In light of these two principles, we can see why Dworkin is not only undeterred in maintaining his liberal egalitarian ethical outlook by the prospect of genetically engineering children, but in fact he finds it positively exhilarating precisely because of maintaining this outlook. Not only does Dworkin think that there is nothing wrong with the "ambition to make the lives of future generations of human beings longer and more full of talent and hence achievement," but he also maintains:

> [If] playing God means struggling to improve our species, bringing into our conscious designs a resolution to improve what God deliberately or nature blindly has evolved over eons, then the first principle of ethical individualism commands that struggle, and its second principle forbids, in the absence of positive evidence of danger, hobbling the scientists and doctors who volunteer to lead it.   (452)

Against the Promethean Ideal

What should we make of the Promethean ideal of playing God, that is, of making a choosing-controlling stance dominant in seeking mastery over the given? I want to begin here to make the case for thinking that the accepting-appreciating stance should be regarded as primary.[15] As mentioned at the outset, I do not deny an important role for the choosing-controlling stance in seeking improvement. However, I contend that it needs to be informed by a proper accepting-appreciating stance. In the sections that follow I will discuss specific limiting virtues that counter the Promethean ideal and give proper expression to an accepting-appreciating stance.

In his book *The Case against Perfection*, Michael Sandel sums up his opposition to the Promethean ideal as expressed in the pursuit of genetic engineering as follows: "The problem with eugenics and genetic

---

[15] My approach here is in contrast not only with those who think the choosing-controlling stance should be dominant but also with those who seek a "balance" between the two existential stances, such as Erik Parens, who seeks a balance between what he calls "the creativity stance" and "the gratitude stance" (*Shaping Our Selves: On Technology, Flourishing, and a Habit of Thinking* [Oxford: Oxford University Press, 2015], ch. 3). See also Parens and Johnston (eds.), *Human Flourishing in an Age of Gene Editing*, where the contributions address the values of control (or shaping) and acceptance, and Part IV is concerned with achieving a balance between them. In my view such a concern with achieving balance is not the right approach, since it suggests that the stances/values are separate from each other, and we need (more or less) an equal amount of both. I do not think the stances should be seen as separate, and I think the accepting-appreciating stance should be seen as more fundamental.

engineering is that they represent the one-sided triumph of willfulness over giftedness, of dominion over reverence, of molding over beholding."[16] In other words, what he is taking issue with here is making the choosing-controlling stance dominant over the accepting-appreciating stance. However, we need to ask: what precisely is wrong with making the choosing-controlling stance dominant? Also, what is the proper relationship between the choosing-controlling stance and the accepting-appreciating stance? I think we need to address the first question in order to make our way to an answer to the second, which will also require us to examine the specific limiting virtues that counter the Promethean ideal. We can only fully work out the proper relationship between the two existential stances, I contend, from the vantage point of the virtuous agent who embodies the limiting virtues and exercises practical wisdom.

It is helpful to begin by exploring what is wrong with making the choosing-controlling stance *completely* dominant such that no place is given to the accepting-appreciating stance. Here I want to build on some remarks from G. A. Cohen in an essay titled "Rescuing Conservatism: A Defense of Existing Value."[17] Although he is well known as a socialist political philosopher, Cohen defends what he calls "small-c conservatism," which is a particular attitude—we might also call it a life-orientation—concerned with conserving "existing value," and which he distinguishes from large-C Conservatism, or political conservatism. I think this can be more aptly called *existential conservatism*, because it is concerned with an accepting-appreciating stance toward what exists, that is, toward the given world.[18] This conservative attitude, Cohen says, "exhibits a bias in favor of retaining what is of value, even in the face of replacing it by something of greater value" (149). He distinguishes between three forms of this conservative attitude: (1) valuing and seeking to preserve that which is intrinsically valuable (i.e., "particular valuing");

---

[16] Michael J. Sandel, *The Case against Perfection: Ethics in an Age of Genetic Engineering* (Cambridge, MA: Belknap Press of Harvard University Press, 2007), 85. Sandel's argument is focused on "enhancement" uses of genetic engineering, but it also applies to creating "designer children" for therapeutic reasons.

[17] G. A. Cohen, "Rescuing Conservatism: A Defense of Existing Value (All Souls version)," in *Finding Oneself in the Other* (Princeton, NJ: Princeton University Press, 2013), 143–74.

[18] See David McPherson, "Existential Conservatism," *Philosophy* 94:3 (2019): 383–407.

(2) valuing and seeking to preserve that which is personally valued (i.e., "personal valuing"); and (3) "accepting the given." I will return in the last section of this chapter to discuss Cohen's accounts of "particular valuing" and "personal valuing," but here I want to focus on what he has to say about "accepting the given."

Cohen says the least about this aspect of his existential conservatism; he mostly sets it aside, "not because it is unimportant," but because he thinks that "it is too deep to explore" within the confines of his essay (152). Indeed, he leaves the key claims in his remarks about it largely undeveloped. However, what he does say is very suggestive, and I think it points us in the right direction. The key claims he makes are the following:

> [Some] things must be accepted as given,…not everything can, or should, be shaped to our aims and requirements; the attitude that goes with seeking to shape everything to our requirements both violates intrinsic value and contradicts our own spiritual requirements…. [The] attitude of universal mastery over everything is repugnant, and, at the limit, insane.   (149)

We should first note here that when Cohen talks about "accepting the given," he means accepting *some* things as given, particularly intrinsically valuable things. With this qualifier in mind, we can identify here two claims to focus on: (1) the stance of universal mastery "violates intrinsic value," and (2) this stance "contradicts our own spiritual requirements."[19] Let us consider each in turn.

---

[19] It should be noted that I am not taking up here a common consequentialist line of argument against Prometheanism, which we might express in terms of the danger of "playing with fire" (to recall Prometheus' act of stealing fire from the gods): the danger is that you might get "burned," that is, it might lead to unintended bad consequences. This occurs especially when we seek to extend our control into areas where we lack knowledge or competence (see John Weckert, "Playing God: What is the Problem?" in Clarke et al. [eds.], *The Ethics of Human Enhancement*, 94, 97–8). This can happen not only with our relation to nature but also with complex human affairs (see the discussion of the politics of imperfection in Chapter 3). We see this "playing with fire" concern in Mary Shelley's *Frankenstein: or, The Modern Prometheus*. We also see something of this concern in Bernard Williams' discussion of how a "sense of restraint in the face of nature" can be grounded in a "Promethean fear," which is "a fear of taking too

I take it that the way in which the stance of universal mastery "violates intrinsic value" is why Cohen finds it repugnant.[20] The key point here is that the stance of seeking mastery over everything (or at least everything in principle) in order to suit our desires does not recognize any proper limits on our will, which comes from rightly appreciating what is of intrinsic value in the given world. If we think that everything is up for grabs for our use—if we think that "the world's mine oyster," as Pistol puts it in Shakespeare's *The Merry Wives of Windsor*—then clearly this will lead to violation of what is of intrinsic value in the given world. For instance, it will lead us to regard other human beings as objects merely to be used for our purposes, rather than as having a special inherent dignity that needs to be respected and that sets limits upon our will, and therefore we will violate this special dignity of human life.

Even where it does not assume the stance of universal mastery, making the choosing-controlling stance dominant can be seen to violate intrinsic value. For instance, Sandel discusses how genetic engineering of children undermines a key aspect of the best kind of parental love, namely, an accepting love that appreciates children as gifts: "To appreciate children as gifts is to accept them as they come, not as objects of our design, or products of our will," that is, it requires an "openness to the unbidden."[21] Some object to the language of "gift" in this context, regarding it as inherently theistic by implying an ultimate Giver of the gift of life beyond the parents.[22] While I think theism does provide a fitting worldview for

lightly or inconsiderably our relations to nature," where instead nature should be understood as "an old, unbounded and potentially dangerous enemy" ("Must a Concern for the Environment Be Centered on Human Beings?," in *Making Sense of Humanity: And Other Philosophical Papers* [Cambridge: Cambridge University Press, 1995], 239). There is something to be said for the consequentialist considerations that are operative here, but they are not the most important considerations in my view. Williams also seems to recognize this when he says that if the Promethean fear is "something that many people deeply feel, then it is something that is likely to be pervasively connected to things that we value, to what gives life the kinds of significance that it has" (239).

[20] Leon Kass also writes of a kind of moral repugnance that "revolts against the excesses of human willfulness, warning us not to transgress what is unspeakably profound" ("The Wisdom of Repugnance," *The New Republic*, June 2, 1997: 20).

[21] Sandel, *The Case against Perfection*, 45; see also 46–50, 86.

[22] See, e.g., Allen Buchanan, *Beyond Humanity?: The Ethics of Biomedical Enhancement* (Oxford: Oxford University Press, 2011), 3. Buchanan also accuses Sandel of using "high-sounding" but "murky" rhetoric in speaking of appreciating life as a gift, which Buchanan thinks is "masquerading" as an argument. This seems itself like a bit of rhetoric, which is also used by other defenders of genetic engineering: for instance, Nick Bostrom says that Leon Kass's

seeing human life as a gift, I also think the language of gift in this context can be employed by theists and non-theists alike (though it is metaphorical in the latter case), since it can be understood as referring to an unmerited good that should be properly accepted and appreciated.[23] The claim then is that genetic engineering takes a wrongful posture toward the unmerited good of children by failing properly to accept and appreciate children as gifts.[24] There are in fact two intrinsic values (or goods) that are violated: the parent-child relationship (in its proper mode as rooted in accepting love) and human life understood as a precious gift. Appreciating children as gifts is important both for avoiding the dehumanization involved in treating children as products and for respecting the equal intrinsic dignity of each human life, which is violated when one lords over a child by treating him or her as a product of one's will.[25] Sandel does acknowledge that in addition to accepting love there is a

appeals to human dignity and the blessings of human finitude in arguing against biotechnological pursuits of immortality amount to "fine phrases and hollow rhetoric" ("The Fable of the Dragon Tyrant," *Journal of Medical Ethics* 31:5 [2005]: 277). One can dismiss a lot of ethical argument in this way, including that of Buchanan and Bostrom. But what is needed is to engage rival stances toward the given world and make a case for a particular vision of the good life and the virtues that are constitutive of it, which is what I am seeking to do here.

[23] See Sandel, *The Case against Perfection*, 92–5, for a similar response to this sort of objection.

[24] In his response to Sandel, Guy Kahane writes:

> When something occurs, whether bidden or unbidden, it can be good, bad or indifferent. If its occurrence is under our control then (so long as we aim at the good) the outcome is more likely to be good than if left to chance. This is why, when we value something, we should try, when possible and permissible, to bring it about. Given this truism, it seems odd to think that, when something matters, we should ever leave it to chance rather than choice.
>
> ("Mastery without Mystery: Why There Is No Promethean Sin in
> Enhancement," *Journal of Applied Philosophy* 28:4 [2011]: 357)

What is missed here is that there are some goods that require appreciation rather than control (for the sake of "enhancement").

[25] Gilbert Meilaender writes:

> In begetting we...give of ourselves and thereby form another who, though other, shares our nature and is equal to us in dignity. If, by contrast, we come to think of the child as a product of our reason and will, we have lost the deepest ground of human equality—and, perhaps as important, have missed the meaning of the human act of love.
>
> (*Bioethics: A Primer for Christians*, 2nd edn
> [Grand Rapids, MI: Eerdmans, 2004], 14)

Kass also regards genetic engineering of children as the despotism of one generation over the next, which "represents a blatant violation of the inner meaning of parent-child relations" ("The Wisdom of Repugnance," 21).

proper place in parental love for a transforming love that seeks the best for one's child.[26] But where we go too far is when we fail to appreciate children as gifts and undermine the proper place of accepting love in the parent-child relationship. The virtuous person knows how to get things right here. In my view, the virtues should be understood as modes of proper responsiveness to what is of intrinsic value, and this will come out in my account of the limiting virtues. In light of this we can see one way that the accepting-appreciating stance needs to be regarded as primary: we first need to appreciate what is of value in order to know how to act or not act. In other words, the accepting-appreciating stance should inform when and how we take up the choosing-controlling stance. I will return in a bit to discuss this point further.

But before doing so we need to consider Cohen's more cryptic claim that the stance of universal mastery "contradicts our spiritual requirements." What exactly are these "spiritual requirements"? He gives the following suggestion at the beginning of his essay:

> [It] is a pregnant moment in the New Testament when Jesus, awaiting his arrest in the Garden of Gethsemane and foreseeing the toils to come, cries out "Oh, Lord, take away this cup," but then corrects himself: "but not my will, Lord, thine." The motif is abandonment of striving, of seeking a better state, and instead going with the flow, as do the lilies of the field, which are at peace with the world, and therefore with themselves.   (143)

It is noteworthy that Cohen is, in fact, an atheist (of Jewish heritage), but nevertheless he finds something spiritually important here. Another way we can put his point about being at peace with the world and therefore with ourselves is to say that one of our key spiritual requirements is to

---

[26] This, of course, includes caring for a child's health. Sandel writes:

> To appreciate children as gifts... is not to be passive in the face of illness or disease. Healing a sick or injured child does not override her natural capacities but permits them to flourish. Although medical treatment intervenes in nature, it does so for the sake of health, and so does not represent a boundless bid for mastery and dominion.
> (*The Case against Perfection*, 46–7)

find a way of being at home in the world despite the evil and suffering it contains. What we are seeking to address here is *the problem of cosmodicy*, which is to say, the problem of justifying life in the world as worthwhile in the face of evil and suffering.

The different versions of the Promethean ideal can also be seen as seeking to address this problem of cosmodicy in their own way. For instance, one might attempt to do so by striving to improve the world as far as possible through science and technology, including genetic engineering, as can be seen with Dworkin when he speaks of "bringing into our conscious designs a resolution to improve what God deliberately or nature blindly has evolved over eons." Here one attempts to "play God" by "correcting the deed" of God, or nature, as Dworkin actually sees it, in creating a world that is, in many respects, out of joint.[27] We also saw a concern with cosmodicy in Nietzsche's attempt to overcome Schopenhauer's no-saying to life through counseling resignation in light of the pervasiveness and ineliminabilty of suffering in human life; against this, Nietzsche maintains that suffering should be affirmed as integral to the process of overcoming resistance and achieving greatness in the expression of the will to power.

We see in both cases an important contrast with Cohen, since he believes that adequately addressing the problem of cosmodicy requires some "abandonment of striving, of seeking a better state." Cohen should not be taken as advocating that we not seek improvement at all, since much of his work in moral and political philosophy is concerned with improvement. But what he is suggesting—rightly, I contend—is that this is not enough; we need to find a way of being at home in the world, which we cannot do through restless striving. In other words, we need a primary place for the accepting-appreciating stance in addressing the problem of cosmodicy.

Part of the issue here is that, given the limits of our existence, we will never actually realize a state of perfection through our striving, and so we need a way of living with imperfection, that is, we need a way of coming to see life in the world as good and worth affirming despite the ill. This

---

[27] "Correcting the deed" is how Ivan Karamazov puts it in Fyodor Dostoevsky's *The Brothers Karamazov*.

requires a stance of contemplation—i.e., appreciative attention—in relation to the given world. Elsewhere Cohen writes: "What one might call *true* religion celebrates life, and the world, and looks for the good in everything."[28] Such an orientation is important for enabling us to become at home in the world. As Michael Hauskeller puts it:

> Seeing the good in what we have got, i.e., appreciating the giftedness of life, helps us feel at home in the world. It creates a bond, connects us to the rest of the world, which then no longer appears hostile and forbidding, an alien place that may perhaps be best described as enemy territory.[29]

By contrast, he says:

> The drive to mastery and the denial of giftedness affirm this enmity. They reinforce an almost Manichaean point of view, according to which it is either "us" (the Promethean, nature-defying, boundary-transgressing, star-reaching human) or "them" (nature as the evil power that prevents us from rising to the stars where we belong).   (74)

The Promethean project of mastery thus contributes to a feeling of existential alienation.

The Promethean project involves a restless pursuit of achievement, as we have seen with Nietzsche in his ideal of "self-overcoming" in the pursuit of great achievements and with Dworkin in his endorsement of the "ambition to make the lives of future generations of human beings longer and more full of talent and hence achievement." However, our achievements themselves are not really complete without our appreciation of them, and so this is another sense in which the accepting-appreciating stance needs to be regarded as primary. We can think here, for example, about the creation story in Genesis, where God creates the world in six days and then *completes* his creation through

---

[28] G. A. Cohen, "One Kind of Spirituality: Come Back, Feuerbach, All Is Forgiven!," in *Finding Oneself in the Other*, 206.

[29] Michael Hauskeller, "Human Enhancement and the Giftedness of Life," *Philosophical Papers* 40:1 (2011): 74.

appreciating it—where he contemplatively beholds it as "very good"—
and resting on the seventh day. The practice of the Sabbath imitates God
in creation: the Sabbath completes our own work through restful appreciation of this work as well as the world in which we live. I will return in
Chapter 4 to discuss the practice of the Sabbath in more detail, and I will
suggest its relevance for everyone, regardless of religious beliefs.

The accepting-appreciating stance is primary in another way, as already
stated: we first need to appreciate what is of value in order to know how
to act or not act. A danger with making the choosing-controlling stance
dominant is that doing so *courts nihilism*, and falling into nihilism—
the condition in which one believes that nothing matters[30]—amounts to
a failure of cosmodicy. I believe something like this thought is behind
Cohen's comment about how the attitude of universal mastery, at the
limit, is "insane." How does making the choosing-controlling stance
dominant court nihilism? Consider the following passage from David
Wiggins:

> If we cannot recognize our own given natures and the natural world as
> setting any limit at all upon the desires that we contemplate taking
> seriously; if we will not listen to the anticipations and suspicions of the
> artefactual conception of human beings that sound in half-forgotten
> moral denunciations of the impulse to see…human beings as things,
> as tools, as bearers of military numerals, as cannon-fodder, or as
> fungibles; if we are not ready to scrutinize with any hesitation or
> perplexity at all the conviction (as passionate as it is groundless, surely,
> for no larger conception is available that could validate it) that every-
> thing in the world is in principle ours or there for the taking; then what
> will befall us? Will a new disquiet assail our desires themselves, in a
> world no less denuded of meaning by our sense of our own omnipo-
> tence than ravaged by our self-righteous insatiability?[31]

---

[30] We might call this *subjective nihilism*, since I am referring to a condition in which a subject
believes that nothing matters; but this is connected to what we can call *objectivism nihilism*, the
state of affairs in which nothing *really* matters, objectively speaking. The belief in objective
nihilism makes for subjective nihilism.

[31] David Wiggins, *Sameness and Substance Renewed* (Cambridge: Cambridge University
Press, 2001), 242; cited in Cohen, "Rescuing Conservatism," 150, n. 16.

In other words, if we come to see everything as up for grabs, where there are no ends or objects of choice of great importance such that they can place constraints on our choices, then this will deflate our sense of the importance of choice; we are left with a disenchanted view of the world and indeed "a new disquiet" assails our desires: why desire *anything*? We see a similar point made by Sandel when he writes:

> There is something appealing, even intoxicating about a vision of human freedom unfettered by the given .... But that vision of freedom is flawed. It threatens to banish our appreciation of life as a gift, and to leave us with nothing to affirm or behold outside our own will.[32]

We can add that when we have nothing to affirm or behold outside of our own will—i.e., nothing considered as being of intrinsic value and demanding our appreciative attention—then we can come to find that we have *nothing to will*. We may, of course, still will on the basis of what we happen to desire, but as meaning-seeking animals we seek to orient our lives toward what is objectively meaningful (or valuable) and worthy of our appreciative attention.[33] We can step back from what we happen to desire and ask why these desires *should matter* to us, and if we are left with nothing to say about their objective importance, then this will deflate our sense of the importance of what we desire, and thus we may find ourselves with nothing to will.[34] Therefore, if we are to ensure avoiding such a debilitating condition, we must regard an appreciative stance toward the given as being more fundamental than any choosing stance. In other words, an appreciative stance toward what is of intrinsic value in the given world provides the necessary background against

---

[32] Sandel, *The Case against Perfection*, 99–100.

[33] I have developed at length the idea of our being the meaning-seeking animal in *Virtue and Meaning*.

[34] Mark Platts speaks of how a lack of recognition of objective value "can be the end for a reflective being like us by being the beginning of a life that is empty, brutish, and long" ("Moral Reality and the End of Desire," in *Reference, Truth and Reality*, ed. Mark Platts [London: Routledge & Kegan Paul, 1980], 80). See also Fiona Ellis, "Desire and the Spiritual Life," in *Spirituality and the Good Life: Philosophical Approaches*, ed. David McPherson (Cambridge: Cambridge University Press, 2017), 84–100.

which genuinely significant choices can be made. As Aurel Kolnai puts it: "*response,* not *fiat,* is the prime gesture of the human person."[35]

We saw Nietzsche and Dworkin acknowledge the lurking threat of nihilism in their positions, as both describe an experience of existential vertigo or disorientation that results from making a choosing-controlling stance toward the given dominant. But both seek to forge ahead undaunted—even suggesting that "the alternative is cowardice"—and they also experience a sense of liberation in being no longer "fettered" by the supposed ethical demands of the given.

As we saw, Nietzsche forges ahead by endorsing an ethic of power, where what is considered good is whatever "heightens the feeling of power in man, the will to power, power itself," and happiness is understood as the "feeling that power is *growing,* that resistance is overcome."

---

[35] Aurel Kolnai, "Privilege and Liberty" (1949), in *Privilege and Liberty and Other Essays in Political Philosophy* (Lanham, MD: Lexington, 1999), 26. Consider also Mary Midgley:

> [A] great deal in the life of each of us is completely out of our power, and our freedom must consist in the way we handle that small but crucial area which does actually come before us for choice. This situation is far more benign than people obsessed with freedom make it sound, because what comes to us by no choice of our own is a gift—a whole world which we could not possibly have made and at which, in spite of all its horrors, we can on the whole only bow our heads in wonder. Moralists able only to think of autonomy, of the active imposition of the will on what is round us, miss the essential values of receptivity, of contemplation, of openness to the splendours of what is not oneself. Our inheritance, both social and natural, is not a shocking intrusion on our privacy and freedom, but a realm for us to live in. Morally speaking, one of the worst aspects of the autonomy-centred Enlightenment attitude has been to denigrate the receptive virtues, to make us so obsessed with giving that we do not know how to receive. As usual, one of these aspects of life does not make much sense without the other.
>
> ("On Not Being Afraid of Natural Sex Differences," in *Feminist Perspectives in Philosophy,* ed. Morwenna Griffiths and Margaret Whitford [London: Macmillan Press, 1988], 39)

In a similar vein, G. K. Chesterton writes:

> Human beings are happy so long as they retain the receptive power and the power of reaction in surprise and gratitude to something outside. So long as they have this they have ... something that is present in childhood and which can still preserve and invigorate manhood. The moment the self within is consciously felt as something superior to any of the gifts that can be brought to it ... there has appeared a sort of self-devouring fastidiousness and a disenchantment in advance, which fulfills all the Tartarean emblems of thirst and of despair.
>
> ("If I Had Only One Sermon to Preach" [1950], in *In Defense of Sanity: The Best Essays of G. K. Chesterton* [San Francisco, CA: Ignatius Press, 2011], 347)

In other words, as I have put it, we court nihilism if we make the choosing-controlling stance dominant.

However, this does not escape the problem identified by Wiggins and Sandel: if there is nothing beyond the will to affirm and appreciate, then there is nothing *worth* willing or desiring. Nietzsche at times does appear to adopt an accepting-appreciating stance toward the given world when he talks about yes-saying to the world, as in the following passage:

> Today everybody permits himself the expression of his wish and his dearest thought: hence I, too, shall say what it is that I wish from myself today, and... what thought shall be for me the reason, warranty, and sweetness of all my life henceforth! I want to learn more and more to see as beautiful what is necessary in things:—then I shall be one of those who make things beautiful. *Amor fati*: let that be my love from henceforth! I do not want to wage war against what is ugly. I do not want to accuse, I do not even want to accuse those who accuse. *Looking away* shall be my only negation! And all in all and on the whole: some day I wish to be only a Yes-sayer![36]

However, for Nietzsche, such yes-saying is, in fact, the ultimate expression of the will to power in the face of life's hardships, rather than a matter of adopting an accepting-appreciating stance toward what is of intrinsic value in the given world; hence he speaks of being "one of those who *make* things beautiful."[37] In other words, yes-saying, for Nietzsche, has to do with the subject side rather than the object side. But again, why

---

[36] Nietzsche, *The Gay Science*, §276.

[37] Consider also:

> To redeem those who lived in the past and to recreate all "it was" into a "thus I willed it"—that alone should I call redemption. Will—that is the name of the liberator and joy-bringer... All "it was" is a fragment, a riddle, a dreadful accident—until the creative will says to it, "But thus I willed it."
>
> (Nietzsche, *Thus Spoke Zarathustra*, II, "On Redemption")

and:

> I have become one who blesses and says Yes; and I have fought long for that and was a fighter that I might one day get my hands free to bless. But this is my blessing: to stand over every single thing as its own heaven.
>
> (Nietzsche, *Thus Spoke Zarathustra*, III, "Before Sunrise"; see also *The Gay Science*, §341)

It should be noted that Nietzsche does not in fact affirm all of life: he does not want to affirm the weak and those who are failures, as he thinks that they give life "a gloomy and questionable aspect" (*The Antichrist*, §7).

affirm anything if there is nothing beyond the will that is *worthy* of affirmation?[38] Moreover, how can we truly make things beautiful if there is nothing inherently beautiful to which we ought to be responsive and which can constrain our will? In short, the problem of nihilism has not, in fact, been overcome.

For his part, Dworkin responds to the threat of nihilism inherent in his position by seeking refuge in what he describes as "a more critical and abstract part of our morality," which he understands in terms of his "ethical individualism," which we saw is committed to the values of individual autonomy and beneficence.[39] But it is not clear that even this can remain intact once we are no longer constrained by the demands of the given. Dworkin's ethical individualism does recognize constraints on choice in virtue of respecting the autonomous choices of others. However, the value of this autonomy is derivative. As I have argued, if there are no ends of choice that are of great importance such that they can place constraints on our choices, then this deflates our sense of the importance of choice (i.e., autonomy). As we have seen, Dworkin also affirms the value of beneficence, though the value of this depends upon there being objects of beneficence that are worthy of our concern, namely, other human beings (the focus here is on human beings, but this is not to deny that nonhuman creatures are also worthy of concern). However, I have argued that genetic engineering of children undermines an appreciation of human life as a gift and involves seeing it as a product of one's will, which violates our dignity as human beings and has the effect of undermining human equality, as we have one generation lording over the next. Furthermore, the view of beneficence as concerned above

---

[38] Simon May believes Nietzsche is still in the "morality game" in so far as he is seeking justification at all (whether in the world or in himself) for the affirmation of life in the face of suffering. However, at times, as in the *amor fati* passage I cited, May thinks that Nietzsche's thought indicates "a struggle to escape theodicy [i.e., cosmodicy] and instead to find an affirmation of life that is an ungrounded joy in life's 'there-ness' or quiddity" ("Why Nietzsche Is Still in the Morality Game," in *Nietzsche's* On the Genealogy of Morality: *A Critical Guide*, ed. Simon May [Cambridge: Cambridge University Press, 2011], 80). On this account, "the affirmer does not take joy in his life as a whole *because* he has seen goodness or beauty in it" (100). But *amor fati* does seem to offer a higher ideal that seeks to justify life for Nietzsche, and it is also not clear what "ungrounded joy" is supposed to be. All intelligible joy, it seems, is grounded in something about ourselves or the world (including others), and it needs to be understood as responsive to what is seen as good or beautiful.

[39] Dworkin, "Playing God," 448–9.

all with promoting better quality of life ultimately leads us to seeing everything in the given world, including humanity, as *replaceable*, as long as this would involve a higher quality of life for whatever transhuman beings there are. There is something very disconcerting about seeing everything in the given world—including humanity—as replaceable by something better. This fails to be properly responsive to what is of value in the given world and in the human condition—e.g., our characteristic human forms of love, moral commitment, aesthetic pursuits, spiritual devotion, and other ways of seeking human fulfillment—and to the way in which this value commands our loyalty. This does not mean we have to think that everything is perfect in the given world and in our humanity. But it does mean that it is important that we find some things of great value that are worth preserving.[40] If nothing is of great value such that it can command our loyalty, then we are again back to the threat of nihilism, that is, that nothing really matters. I will return in the final section of this chapter to discuss further this issue of loyalty to the given.

For now, I want to turn to discuss the related limiting virtues of humility and reverence. As previously mentioned, to work out the proper relationship between the choosing-controlling stance and the accepting-appreciating stance we need to examine the limiting virtues, which are concerned with achieving this proper relationship. Indeed, I have contended that we can only fully work out a proper relationship between these existential stances from vantage point of the virtuous agent who embodies the limiting virtues and exercises practical wisdom.

Humility and Reverence

As mentioned in the Introduction, humility can be regarded as the master limiting virtue: it ensures that we recognize and live out our proper place in the scheme of things, and it is concerned, among other

---

[40] Buchanan thinks that opponents of biomedical enhancement overlook the problematic features of our human nature (e.g., selfish tendencies) that have resulted from the evolutionary process (see *Beyond Humanity?*, ch. 4). There is no need to deny there are such features, but this does not gainsay that there is much that is of great value that is integrally bound up with our human condition that we should think is worth preserving.

things, with reining in the tendency to "play God." One traditional way of framing the virtue of humility is to say that it involves properly acknowledging that we are situated between the beasts (i.e., non-rational animals) and the divine, and it ensures that we do not think of ourselves as on a par with the divine. To think of ourselves as on a par with the divine is a form of hubris, or improper pride, which is the vice opposed to the virtue of humility. However, it is also important that we do not think of ourselves as on a par with the beasts, and for this reason humility needs to be paired with proper pride, or magnanimity, which recognizes the special dignity of human life, and our capacity for and realization of greatness and nobility.[41] Sometimes this has been framed in terms of recognizing that, though we are not God, we are made in the image of God (as in Judaism and Christianity) or that we have an element of the divine within us (as Aristotle puts it). The virtue of humility should not be seen then as involving self-abasement that is incompatible with our special dignity as human beings. Indeed, because humility gives proper recognition to our humanity and its place in the scheme of things, it plays a crucial role in the achievement of human virtue. As Martha Nussbaum writes: "Human limits structure the human excellences, and give excellent action its significance.... [Hubris is] the failure to comprehend what sort of life one has actually got, the failure to live within its limits."[42]

It should be noted that hubris, or improper pride, is found not only in attempts to see ourselves as on a par with the divine; it is found in any form of thinking too highly of ourselves. In light of this, Robert C. Roberts and W. Scott Cleveland have put forward an account of the virtue of humility according to which it is regarded as an "intelligent lack of concern for self-importance," which is to say, "the intelligent absence

---

[41] My account here has significant resemblance to that of St. Thomas Aquinas, who says that with respect to our aspirations toward goods that are difficult to attain we need two virtues: "one, to temper and restrain the mind, lest it tend to high things immoderately; and this belongs to the virtue of humility; and another to strengthen the mind against despair, and urge it on to the pursuit of great things according to right reason; and this is magnanimity" (*Summa Theologiae* II–II, q. 161, a. 1, trans. Fathers of the English Dominican Province). Aristotle's well-known account of magnanimity goes wrong, I believe, because it lacks the connection with proper humility, and so it becomes improper pride (see *Nicomachean Ethics*, IV.3).

[42] Martha C. Nussbaum, "Transcending Humanity," in *Love's Knowledge: Essays on Philosophy and Literature* (Oxford: Oxford University Press, 1990), 378, 381.

of the vices of pride," such as arrogance, vanity, snobbery, hyperautonomy (i.e., aspiring to be self-made), self-righteousness, haughtiness, grandiosity, domination, and conceit.[43] I think this is on the right track; however, it defines humility too negatively. The virtue of humility does not just involve reining in the tendency to give too much place to a choosing-controlling stance, which goes along with an inflated sense of self-importance, but also involves giving a proper role to the accepting-appreciating stance toward the given; it recognizes that some things must be accepted and appreciated *as given*, and not subject to human control or manipulation. As Sandel puts it: "An appreciation of the giftedness of life constrains the Promethean project and conduces to a certain humility."[44]

In addition to appreciating the giftedness of life, it is an important part of the virtue of humility to accept and appreciate the dependent nature of our existence; we depend not only upon others and the natural world for our existence but also upon values not of our own making for living meaningful lives. The virtue of humility thus recognizes, as we saw Kolnai put it, that "*response*, not *fiat*, is the prime gesture of the human person." Further, giving a proper place to an accepting-appreciating stance toward the given means properly acknowledging human limits, including not only natural and personal limitations—as is done in limits-owning accounts of humility[45]—but also moral limitations. The virtue of humility is concerned with giving proper recognition to self-transcending sources of value that place constraints on our will and thereby help to define our proper place in the scheme of things.

This means that the limiting virtue of humility needs to be understood as closely connected with another limiting virtue: reverence. We could also speak of the importance of respect here as a limiting virtue, but to simplify the discussion I want to focus on the virtue of reverence, which

---

[43] Robert C. Roberts and W. Scott Cleveland, "Humility from a Philosophical Point of View," in *Handbook of Humility: Theory, Research, and Applications*, ed. Everett L. Worthington et al. (New York: Routledge, 2017), 33–7.

[44] Sandel, *The Case against Perfection*, 27; see also 45–6, 86.

[45] See Nancy E. Snow, "Humility," *Journal of Value Inquiry* 29:2 (1995): 203–16; Dennis Whitcomb, Heather Battaly, Jason Baehr, and Daniel Howard-Snyder, "Intellectual Humility: Owning Our Limitations," *Philosophy and Phenomenological Research* 94:3 (2017): 509–39; and Nathan Ballantyne, *Knowing Our Limits* (Oxford: Oxford University Press, 2019).

can be understood as a heightened form of respect, and so as recognizing particularly strong constraints on our will. Hence we saw Sandel appeal to reverence as the contrast to the stance of dominion when he objected to the way in which genetic engineering wrongly prioritizes "dominion over reverence." As I understand it, the virtue of reverence is concerned with being properly responsive, through reverential attitudes and behavior, to that which is reverence-worthy and which places strong constraints on our will. The reverence-worthy can be understood as that which has special dignity (hence the feeling of reverence can be understood as a heightened feeling of respect), or it might also be described as that which is sacred or holy. One example of something that is reverence-worthy is human life, due to its ethical and spiritual capacities, and thus the virtue of reverence involves proper responsiveness to the special dignity or sanctity of human life, which includes recognizing the limits that other human beings place on our will. For the theist, God is worthy of reverence to an even greater extent (indeed, God is worthy of worship), in virtue of God's holiness, and so places limits on our will, including that we are not to play God. The virtue of reverence can, in fact, be seen as synonymous with the virtue of piety, which is often characterized in terms of showing reverence for the sources of our existence—such as God (religious piety) and our parents (filial piety)—but which can be understood as being concerned with showing reverence for whatever is reverence-worthy and with recognizing the constraints that this places on our will.[46]

---

[46] Cora Diamond writes:

> The notions of piety and impiety are complex...One part of the notion of piety is the idea that we should treat the natural order of things with respect and awe. Another element is the idea of respect and gratitude as due to the sources of our life, which may be conceived to be God or the gods, our parents and ancestors, and our country. Another part of the notion of piety is the idea that actions that violate piety are properly regarded as outrageous or shocking. This outrage may manifest a sense that a wholly wrongful posture has been exhibited, a kind of will to dominate the natural order, a refusal to accept limitation, a challenge to God's sovereignty or to honored and honorable traditions that should be taken as sacred.
>
> ("The Problem of Impiety," in *Spirituality and the Good Life: Philosophical Approaches*, ed. David McPherson [Cambridge: Cambridge University Press, 2017], 31)

A key difference between humility and reverence is that humility is self-referential in a way that reverence is not. As I have said, the virtue of humility is concerned with recognizing and living out our proper place in the scheme of things, whereas the virtue of reverence is concerned with being properly responsive, through reverential attitudes and behavior, to that which is reverence-worthy. It is in part because that which is reverence-worthy helps to define our proper place in the scheme of things—through placing constraints on our will—that humility and reverence are closely connected limiting virtues. But it is also important to note that the virtue of humility helps to facilitate the virtue of reverence because of recognizing the importance of an accepting-appreciating stance toward the given world.

It should be acknowledged that both humility and reverence (or piety) have religious connotations, and for this reason some might object to them. Indeed, I have noted that a traditional way of framing the concern of the virtue of humility is in terms of properly situating ourselves between the beasts and the divine. Can one reject this framing and still maintain that humility and reverence are virtues? I think so, as long as one acknowledges that there are things that are reverence-worthy and place constraints on one's will, such as human life and its sources, and it seems that many nonreligious people do in fact acknowledge a place for the reverence-worthy in their lives. If there are no virtues of humility and reverence, then we are free to take up the Promethean project of playing God. But then we are also back to the threat of nihilism: if there are no ends of choice that are of great importance such that they can place constraints on our choices, then this deflates our sense of the importance of choice. If we are to ensure that we avoid this debilitating condition, then an accepting-appreciating stance toward the given world needs to be regarded as primary, and humility and reverence need to be regarded as virtues.

The close connection between humility and reverence can be seen in how sometimes these virtues are confused, as we see in Paul Woodruff's characterization of the virtue of reverence. He writes:

Reverence begins in a deep understanding of human limitations; from this grows the capacity to be in awe of whatever we believe lies outside our control—God, truth, justice, nature, even death…. Simply put, reverence is the virtue that keeps human beings from trying to act like gods.[47]

Understanding human limitations is, in fact, part of the virtue of humility, as is keeping us from "trying to act like gods." Reverence as a virtue, on the other hand, is not just a "*capacity* to be in awe"—which all human beings have in virtue of being human—but rather it is a *disposition* for appropriately feeling and expressing awe and reverence for that which is worthy of awe and reverence. Furthermore, the object of reverence is not adequately characterized by saying that it is "whatever we believe lies outside of our control." There are things outside of our control that are not worthy of reverence (e.g., the weather), and there are things that may be within our control (to some degree) that are worthy of reverence (e.g., another human being), but because of being reverence-worthy *ought not to be controlled*. In other words, the virtue of reverence involves a disposition to feel and express reverence for that which is reverence-worthy, and it recognizes limits upon our will with the knowledge that in some cases we can (and some people do) transgress these limits.

Woodruff misunderstands reverence at least in part because he misunderstands humility, which he unnecessarily characterizes in a solely negative light, as being essentially a vice. He writes:

Reverence is not humility. The opposite of reverence is hubris—which is always a bad thing—but the opposite of humility is pride, and pride can be a good thing…. Humility smacks of obedience to authority, sometimes even of obsequiousness. But a reverent soul can stand up to authority and is never obsequious.   (61)

The opposite of reverence, I contend, is not hubris, but rather it is irreverence. I have also acknowledged that there is a proper form of pride, but clearly we also commonly speak of an improper form of pride—as I have done—which is equivalent to hubris, and it is this improper pride that is the

---

[47] Paul Woodruff, *Reverence: Renewing a Forgotten Virtue*, 2nd edn (Oxford: Oxford University Press, 2014), 1.

opposite of humility. Likewise, there is an improper kind of "humility" that takes the form of self-abasement (which is the kind of "humility" that Woodruff is discussing), but this is to fail to attain proper humility, which I have suggested needs to be paired with a proper form of pride. We should not neglect proper humility because it has its own distinct focus, even while it is closely connected with reverence: the virtue of humility seeks to recognize and live out our proper place in the scheme of things, and the virtue of reverence helps to define what this proper place is in virtue of being properly responsive to self-transcending sources of value that place constraints on our will. In short, the virtue of reverence is necessary for achieving the aim of the virtue of humility.

Contentment and Gratitude

Let us now consider another important limiting virtue that helps us to realize a proper relationship between the choosing-controlling stance and the accepting-appreciating stance toward the given, namely, the virtue of contentment.

It should be noted that contentment is often understood as equivalent to happiness, where one has achieved the satisfaction of one's desires. We have already seen Nietzsche challenge the value of such contentment when he describes the happiness of "the last man"—the epitome of modern leveling culture—as "wretched contentment," whereas he believes one's "happiness ought to justify existence." Such contentment is essentially understood in hedonistic terms ("One has one's little pleasure for the day and one's little pleasure for the night") and as disconnected from any higher aims ("'What is love? What is creation? What is longing?...' thus asks the last man, and he blinks"). What we have here, in fact, is a depiction of *dehumanized* humanity in line with what is depicted in Aldous Huxley's dystopian futuristic novel *Brave New World*. But surely there is a place we can occupy between such dehumanized humanity that seems subhuman and Nietzsche's transhumanist ideal of the overman, where one engages in restless self-overcoming and happiness involves the "feeling that power is *growing*, that resistance is overcome." I have argued that Nietzsche's account of happiness is, in fact, not able to "justify existence" because of lacking a proper accepting-appreciating stance (though his account of yes-saying to the world could be developed in this direction) and that we need to find a way to be at home in the world (as far as possible) despite its imperfections. My contention is that contentment understood as a virtue is important for this task of becoming at home in the world.

The virtue of contentment is the virtue of knowing when enough is enough, of being properly satisfied, of not wanting more than is needed for a good life. It does not deny that we ought to seek improvement in many ways—we should seek growth in the virtues, and there is often nothing wrong and indeed much that is good in seeking to better the world in which we live—but it acknowledges that we need to find a way to be happy and at home in the world amidst imperfection. It involves a *good enough* judgment. In her essay "On Being Content with Imperfection," Cheshire Calhoun provides an account of contentment as a virtue that responds to imperfect conditions—which, of course, characterizes the human condition—and involves "a disposition to appreciate the goods in one's present condition and to use expectation frames that enable such appreciation."[48] Indeed, she describes it as a "virtue of appreciation" (344), that is, it is a virtue that gives a proper place to an accepting-appreciating stance toward the given world. Calhoun brings out the importance of contentment by identifying what is wrong with being someone who is disposed to being discontent:

[There] seems something deficient in being disposed to focus on what is flawed, inadequate, or disappointing.... The problem is that the discontent seem deficient in a capacity for grateful appreciation of what is good even in imperfect circumstances, intolerant of imperfection, or narrowly focused on their own welfare... Most fundamentally, the chronically discontent may strike us as having gone wrong not by exaggerating how bad their circumstances are, but in having misplaced, or at least unnecessarily high, expectations about how good their circumstances must be in order to be good enough to be content with.

(329–30)

---

[48] Cheshire Calhoun, "On Being Content with Imperfection," *Ethics* 127:2 (2017): 327.

It is certainly possible that one's circumstances can genuinely fail to be good enough, and thus there is a proper place for discontent in human life. Furthermore, one can be discontent with certain aspects of one's life, but be content overall with life, feeling that it is good enough. What matters for the virtue of contentment is that we have a disposition to appreciate what is good in the given world and that we have "expectation frames that enable such appreciation."

One way to bring out the idea of "expectation frames" here is with the common saying that we can see the "glass"—i.e., our life, our present circumstances, the given world, etc.—as either half-full or half-empty. This connects up with the remark from Cohen that I cited earlier, where he says: "What one might call *true* religion celebrates life, and the world, and looks for the good in everything." The wider context for this remark is a discussion by Cohen of his own discovery of the deeper wisdom found in the common saying about how we can see the glass as either half-full or half-empty, which he had previously considered banal. He writes:

> When we say [the glass is] half-full, we celebrate what we have. Instead of measuring what we have by some ideal, which leads to half-empty, we measure it from base 0, and then all goods are boons.... [What's] good is, often anyway, good enough, that it [is] wiser, often, to satisfice than to maximize.... [This is] because your life is at least half-empty if you measure what you've got by what you might have and more than half-full if you take every boon as a boon, as part of what some think of as God's bounty. What one might call *true* religion celebrates life, and the world, and looks for the good in everything.[49]

The idea that it is often wiser "to satisfice than to maximize" is of central importance for anyone who recognizes the limiting virtue of contentment. If one is always seeking to maximize, then one will be perpetually discontent, since there is always more to be sought. As discussed, it is one of our key spiritual requirements that we find a way of being at home in

---

[49] Cohen, "One Kind of Spirituality," 206.

the world despite its imperfections, including our own human limitations.[50]

What I want to develop further is Cohen's idea that we should measure everything from "base 0" because then "all goods are boons," or we might say "gifts," to recall the earlier discussion of appreciating the giftedness of life. What does it mean to measure everything from "base 0"? We should turn here to G. K. Chesterton, since the central theme of his work is the importance of giving appreciative attention to the sheer gratuitousness of existence, which he also connects with contentment, as we see when he writes: "The aim of life is appreciation; there is no sense in not appreciating things; and there is no sense in having more of them if you have less appreciation of them."[51] In one of his best-known books, *Orthodoxy*, Chesterton brings out the sheer gratuitousness of existence when discussing one of his favorite books from his boyhood, *Robinson Crusoe*. He says that the best part of the novel is the list of things that were saved from the wreck of Crusoe's ship, and he comments:

> It is a good exercise, in empty or ugly hours of the day, to look at anything, the coal-scuttle or the book-case, and think how happy one could be to have brought it out of the sinking ship on to the solitary island. But it is a better exercise still to remember how all things have had this hair-breadth escape: everything has been saved from a wreck [i.e., non-existence] .... Men spoke much in my boyhood of restricted

---

[50] David Wiggins, who like Cohen is an atheist, also endorses a certain religious (or quasi-religious) frame of mind connected with contentment in his discussion of the ancient Roman virtue of *religio*:

> The thing that matters... is not the doctrinal content of *religio* but the frame of mind—the propensity to cultivate contentment in such good things as the world furnishes, watchfulness on behalf of those good things, and awareness of human limitations such as our ignorance and our imperfect mastery of the natural and social processes we live in the midst of. At this point in the development of the earth, what could be a more reasonable frame of mind?
>
> ("Nature, Respect for Nature, and the Human Scale of Values,"
> *Proceedings of the Aristotelian Society* 100 [2000]: 30)

[51] G. K. Chesterton, *The Autobiography of G. K. Chesterton* (San Francisco, CA: Ignatius Press, 2006 [1936]), 327. Elsewhere he also writes: "[The] proper form of thanks [for the goodness of the world] is some form of humility and restraint: we should thank God for beer and Burgundy by not drinking too much of them" (*Orthodoxy* [1908], in *The Collected Works of G. K. Chesterton, Vol. I* [San Francisco, CA: Ignatius Press, 1986], 268).

or ruined men of genius: and it was common to say that many a man was a Great Might-Have-Been. To me it is a more solid and startling fact that any man in the street is a Great Might-Not-Have-Been.[52]

In light of this passage, we can understand the idea of measuring everything from base 0 in terms of an awe-filled recognition of the fact that we might never have been, that we have been saved from the wreck of non-existence, which provides an expectation frame that enables us to appreciate every good of the given world as a gift or boon.

To regard non-existence as a "wreck" from which we have been saved is, of course, to take a stance on the key issue at stake in the problem of cosmodicy: that it is good to be here rather than never to have been, despite the imperfections of the given world. Indeed, Chesterton goes even further in seeing something truly great in human life when he refers to each of us as a "Great Might-Not-Have-Been," despite our human imperfections. If contentment is to be regarded as a genuine virtue, then I think this does require that we find our way to an affirmation of the world as a whole and human life in particular as being very good indeed, and certainly good enough despite the imperfections.

The crucial point regarding the virtue of contentment, therefore, concerns our *orientation* toward the given world. We can state what is at issue here in terms of the following question: is our basic outlook on the world as it is centered on affirmation or repudiation, yes-saying or no-saying? These are not mutually exclusive options, but the question concerns the emphasis of a particular outlook. The content person who is at home in the given world—in contrast to the discontent person who is alienated from it—is fundamentally affirmative: there is an emphasis that the world, as it is and despite the evils and imperfections it contains, is meaningful and worth affirming, that is, the given world as a whole is good and is a source of joy and fulfillment, even if not everything about it is good and even if there are ways, whether minor or major, that it should be made better. The person with the virtue of contentment first seeks to count his or her blessings, to take stock of what is good about the given

---

[52] Chesterton, *Orthodoxy*, 267.

world, before figuring out how to make it better. In other words, there is an emphasis on gratitude or appreciation.

The kind of gratitude that is at issue here we can call "existential gratitude," which is gratitude for the sheer gratuitousness of existence. Both Calhoun and Cohen want to affirm this sort of gratitude. However, one might question whether they can do so without belief in God, since gratitude seems to involve a "to-for" structure: we are thankful *to* someone *for* some gratuitous good.[53] What existential gratitude requires then is *someone* (namely, God) who is the ultimate source of the unmerited good of one's existence and of all existence and *to whom* one is thankful, which includes a feeling of indebtedness and a desire to pay back the debt in some way. Calhoun puts forward the idea of "propositional gratitude," which is "gratitude that some good state of affairs occurred and is contrasted with gratitude to some agent for the benefit the agent has conferred."[54] I think this is, in fact, better understood in terms of appreciation rather than gratitude (which is a specific kind of appreciation). But, following my earlier remarks about appreciating life as a gift, we can allow a metaphorical extension of existential gratitude for those who want to speak about life as a "gift" as a way of expressing their appreciation for the unmerited good of life, even though, strictly speaking, they do not believe that life is a gift in any ultimate sense, but rather it is good luck. Cohen wants to avoid saying that talking about the good things of life as a "blessing" as opposed to just good luck is "*merely* metaphorical." He writes:

> I do not say, of course, that to speak nontheistically, of being blessed is not to use a metaphor.... But to suggest that, if God is not affirmed, then the language is *merely* metaphorical is ... to imply that the metaphor conveys nothing that could not be put, as it were, in fully prosaic terms [such as saying that one is just lucky]. And that is what I am inclined to deny.[55]

---

[53] See Robert C. Roberts, "The Blessings of Gratitude: A Conceptual Analysis," in *The Psychology of Gratitude*, ed. Robert A. Emmons and Michael E. McCullough (Oxford: Oxford University Press, 2004), 58–78.

[54] Calhoun, "On Being Content with Imperfection," 332, n. 17.

[55] Cohen, "One Kind of Spirituality," 202.

Cohen is making an interesting point here in suggesting that there can be aspects of a non-theist's experience that can only be adequately expressed by deploying language that is at home in a theistic worldview. However, it does seem that there is some loss of meaning if one cannot fully inhabit the non-metaphorical theistic way of speaking about life as a gift.[56]

As concerns the virtue of contentment, what is most important is that we cultivate a disposition toward appreciation of what is good in the given world through which we can come to affirm the world as a whole as good, as good enough despite its imperfections, and thereby become at home as far as possible within it.

Loyalty to the Given

In this task of affirmation and becoming at home in the world it is important to emphasize that we should not see ourselves as being in the position of a detached observer, but rather we should see ourselves as engaged participants in the stuff of life. Consider again Chesterton:

The assumption of [the alternative between the optimist and the pessimist] is that a man criticises this world as if he were house-hunting, as if he were being shown over a new suite of apartments.... But no man is in that position [with respect to the world]. A man belongs to this world before he begins to ask if it is nice to belong to it. He has fought for the flag, and often won heroic victories for the flag long before he has ever enlisted. To put shortly what seems the essential matter, he has a loyalty long before he has any admiration.... [It] seems to me that our attitude towards life can be better expressed in terms of a kind of military loyalty than in terms of criticism and approval. My acceptance of the universe is not optimism, it is more like patriotism. It is a matter of primary loyalty.[57]

---

[56] I discuss this issue of existential gratitude more extensively in *Virtue and Meaning*, 182–6.
[57] Chesterton, *Orthodoxy*, 269–70.

The kind of loyalty at issue here we can call *loyalty to the given*. The virtue of loyalty can be regarded as a limiting virtue because it expresses proper partiality, and thus it places limits on the extent of our attachments. Loyalty involves a binding attachment to someone or something that is maintained through thick and thin. We tend to think of loyalty as an other-regarding virtue that applies to other persons, but it can also apply to any valuable thing, and indeed, as the passage from Chesterton suggests, it can apply to the given world as a whole, in contrast to other possible worlds. We can say that this provides the wider context in which more particular loyalties find their place, and hence it is the "primary loyalty."

In order to fill out and defend the idea that the given world commands our loyalty I want to return to Cohen's defense of existing value. I have already discussed what he has to say about the importance of accepting the given, but we should also consider what he has to say about two distinct modes of valuing, namely, particular valuing and personal valuing.

In the case of *particular valuing*, Cohen says:

a person values something as the particular valuable thing that it is, and not merely for the value that resides in it…. [Even] though the particular indeed gets its value, in the first instance, from the intrinsic value that it has, our valuing of it, the particular, is not merely a valuing of the intrinsic value that it has, but also a valuing of *it*, the particular itself.[58]

He speaks of valuing *something* here, but we can also speak of valuing and loving *someone*, for example, a sweetheart or a friend. We do not simply love and value the beloved because of the beloved's intrinsically valuable qualities; we also love and value the beloved *for the beloved's own sake*, as the particular person she or he is. If we found someone else who had similar or even certain better qualities, we would not say that this other person would do just as well or better. The beloved is

---

[58] Cohen, "Rescuing Conservatism," 148.

irreplaceable. We can also say that a particular charming old building or a plot of land or a cultural way of life is irreplaceable. Cohen remarks:

> [We] devalue the valuable things we have if we keep them only so long as nothing even slightly more valuable comes along. *Valuable things command a certain loyalty.* If an existing thing has intrinsic value, then we have reason to regret its destruction as such, a reason that we would not have if we cared only about the value that the thing carries or instantiates.   (153; my emphasis)

In light of the discussion of existential gratitude in the preceding section, I would add that existing valuable things (or persons) can rightly be seen to command our loyalty as a kind of duty of gratitude for the unmerited given good (or giftedness) of the world. Such a mindset, concerned as it is with the conservation of *what* has value and not merely with the conservation of value, is fundamentally opposed to a maximizing mindset that seeks to maximize value at the expense of valuable things, which are therefore regarded as dispensable. As mentioned when discussing Dworkin on genetic engineering, there is something very disconcerting about seeing everything in the given world, including humanity, as replaceable by something better: this fails to be properly responsive to what is of value in the given world and to the way in which it commands our loyalty. If nothing is of great value such that it can command our loyalty, then we are back to the threat of nihilism, that is, that nothing really matters.[59]

---

[59] Jonathan Pugh, Guy Kahane, and Julian Savulescu claim that while Cohen's conservatism can support some restrictions on uses of human enhancement technologies, it cannot rule out that there are proper uses of them. For one thing, they note that Cohen himself allows that his conservative bias in favor of existing value can be overridden by considerations of justice. When justice is understood according to Cohen's luck egalitarian position that seeks to correct the influence of brute luck in human affairs, then it can support some forms of genetic engineering ("Cohen's Conservatism and Human Enhancement," *The Journal of Ethics* 17:4 [2013]: 343). In Chapter 3 I will return to discuss Cohen's luck egalitarianism, which I do think is in tension with his existential conservatism. I argue that luck egalitarianism should be rejected because of its problematic relationship to the given world, and we should instead adopt a sufficientarian position, where what is important is that everyone has enough.

Pugh et al. also maintain that Cohen's conservatism could be used to support forms of enhancement that conserve existing value, such as enhancements that extend the lifespan and ward off the decline of our normal cognitive capacities (344–5). But the unlimited pursuit of these enhancements would change the human condition as we know it and undermine much of

Once we experience such loyalty, then we are also in the realm of what Cohen calls *personal valuing*, which is where "a person values something because of the special relation of the thing to that person" (148), namely, because of our *attachment* to it. Such a mode of valuing is also resistant to the maximizing mindset. For instance, we can think of an attachment to a house: perhaps we could find what is in many respects a better house, but we do not want to move there because it would mean giving up *our home*, and here "home" expresses our attachment to this particular house and our sense of belonging there. (Of course, sometimes the requirements of work or an expanding family may necessitate finding a new house, and perhaps it is even a better house in many respects, but still if a particular house is a home for us, then we would leave it mournfully.) We can say something similar about a particular plot of land, culture, town, country, friend, and so on. Cohen writes: "We are attached to particular things [or persons] because we need to *belong* to something [or someone], and we therefore need some things [or persons] to belong to us" (168). Cohen does not fill out this point in detail, but we should add that such a need for belonging also takes an existential form in what I have described earlier as a need to be at home in the world.

what we find valuable in it. In his account of "the virtues of mortality" and the place of procreation and childrearing in the human good, Leon Kass writes:

> To be mortal means that it is possible to give one's life, not only in one moment, say, on the field of battle, but also in the many other ways in which we are able in action to rise above attachment to survival. Through moral courage, endurance, greatness of soul, generosity, devotion to justice—in acts great and small—we rise above our mere creatureliness, spending the precious coinage of the time of our lives for the sake of the noble and the good and the holy.
>
> ("*L'Chaim* and Its Limits: Why Not Immorality?," in *Life, Liberty, and the Defense of Dignity: The Challenge for Bioethics* [San Francisco, CA: Encounter, 2002], 267–8)

Also recall Nussbaum's remark on how human limits "structure the human excellences." Additionally, if the enhancements proposed by Pugh et al. involve creating "designer children" then they would also be subject to the objection of treating children as products of one's will rather than as gifts.

Finally, Pugh et al. claim that the use of enhancement technologies can be expressive of a loyalty to our humanity, writing: "it seems plausible to claim that part of what is so valuable about human nature is our capacity and readiness to improve ourselves. The enhancement project then may be regarded not as an expression of self-hatred, but rather of self-realisation" (352). However, this neglects the integral role of the accepting-appreciating stance for living well as human beings and the importance of seeing life as a gift.

We can see this expressed in Chesterton's account of loyalty to the given world, where, because one has already fought for "the flag of the world," one "belongs to this world before he begins to ask if it is nice to belong to it," and one has a "loyalty long before he has any admiration." However, I do not think such loyalty should ultimately be understood as disconnected from appreciation of what is of value in the given world; it needs to be understood as well founded, and indeed as demanded by existing value; otherwise, it is not clear why it would be wrong to defect from the given world. We can see that Chesterton agrees with this point in his account of the wrong of suicide—the ultimate kind of defection— which follows from his account of loyalty to the given world; he writes:

> Not only is suicide a sin, it is the sin. It is the ultimate and absolute evil, . . . the refusal to take the oath of loyalty to life. The man who kills a man, kills a man. The man who kills himself, kills all men; as far as he is concerned he wipes out the world . . . . He defiles every flower by refusing to live for its sake. There is not a tiny creature in the cosmos at whom his death is not a sneer.[60]

I cannot enter into a discussion here about the ethics of suicide (in Chapter 2 I will defend the absolute prohibition against intentionally taking innocent human life), but the idea that I would like to take away from this passage is that the given world can rightly be seen as placing demands upon us for loyalty, and we can fail to be properly responsive to

---

[60] Chesterton, *Orthodoxy*, 276. Compare Ludwig Wittgenstein: "If suicide is allowed, then everything is allowed. If anything is not allowed, then suicide is not allowed. This throws a light on the nature of ethics, for suicide is, so to speak, the elementary sin" (*Notebooks, 1914–1916*, 2nd edn, trans. G. E. M. Anscombe [Chicago: University of Chicago Press, 1961], 91). Philip R. Shields comments:

> [If] suicide is permitted that could only mean that there are no limitations; nothing is required of us, so "everything is allowed." However, for Wittgenstein . . . there *are* limitations, a world of them, beginning with logical form, and in committing suicide a person manages, as it were, to transgress all these limitations at once. This reading gives some depth to McGuinness's gloss: ". . . I think [Wittgenstein's] thought is that [suicide] is the ultimate form of non-acceptance of what happens." . . . The refusal to accept anything as given is a refusal to accept any limitations, and this involves the height of hubris or willfulness.
>
> (*Logic and Sin in the Writings of Ludwig Wittgenstein* [Chicago: University of Chicago Press, 1993], 66)

existing value in being disloyal to it by refusing to belong to the given world. In other words, such loyalty needs to be understood as part of a proper accepting-appreciating stance toward the given world in so far as we believe it contains things and persons that are worthy of our appreciative attention.

Now that we have considered the proper place of limits in human life at the most general level of existential limits, where we are concerned with our stance in relation to the given world, we are in a position to explore the importance of limits in more specific domains.

2

Moral Limits

In this chapter I want to discuss moral limits. In a broad sense of "moral," which is concerned with how we ought to live our lives, each of the limits that I explore in this book can be regarded as involving moral limits. But in this chapter I am focusing on traditional areas of concern for moral philosophers. In the first section, I will discuss the issue of character formation and how this properly begins from learning to acknowledge *restraints* on our desires, which enables us to become the best of all animals, rather than the worst, that is, it enables us to become fully human (in the sense of realizing a normative ideal for humanity). I will focus here on the importance of cultivating the limiting virtues of reverence and moderation. In the next two sections I will be primarily concerned to argue against consequentialist forms of ethics. I will argue in favor of absolute prohibitions by appealing again to the limiting virtue of reverence. I will also critique other proposed bases for such absolute limits on our willing. I will then turn to consider what we positively owe others in terms of assistance. While I think we should have concern for all human beings qua human beings, I will discuss how the limiting virtues of neighborliness (a form of human solidarity that recognizes the moral significance of proximity) and loyalty place constraints on what we can be asked to do on behalf of others.

Character Formation: Beginning with Restraint, Cultivating Reverence and Moderation

At the beginning of the *Politics*, Aristotle remarks: "as a human being is the best of the animals when perfected, so when separated from law and justice he is worst of all . . . . Hence he is the most unrestrained and most

The Virtues of Limits. David McPherson. Oxford University Press. © David McPherson 2022.
DOI: 10.1093/oso/9780192848536.003.0003

savage of animals when he lacks virtue" (I.2, 1253a32–7).[1] Building on this passage, Leon Kass states well what I believe is the basic human predicament to which the process of character formation should seek to respond:

> Man's rationality lies behind his potential savagery, no less than his excellence, because it lies behind his broad, open, and undetermined appetites.... The capacity to think almost everything makes possible the capacity to do almost everything, by means of the desire to appropriate or control whatever is alien.... Possessed of indeterminate and potentially unlimited appetites, willing and able to appropriate... nearly anything in the...world for his own use and satisfaction, man stands in the world not only as its most appreciative beholder but also as its potential tyrant. If he is not to become the worst of the animals, he must be restrained by law and justice. And if he is to become the best of animals, he must be perfected by rearing in customs that bring out and complete what is best in his nature. As with the ethical in general, the primary need is for restriction, for nay-saying; if too much latitude is the problem, then a prohibition is the beginning of the civilizing process. Holding down is the precondition of drawing up. In the first place, man's protean and indeterminate appetites need to be delimited and constrained. But custom will not only restrict and constrain. It will also embellish and dignify.[2]

Here we see Kass identifying expressions of the two fundamental existential stances that I discussed in Chapter 1—the choosing-controlling stance and the accepting-appreciating stance—when he says that "man stands in the world not only as its most appreciative beholder but also as its potential tyrant." But he adds a crucial element to the picture of our human form of life that I have developed so far, which is the acknowledgment of our "broad, open, and undetermined appetites," and which we see is connected to our being rational animals. The key initial task of character formation therefore is to *delimit* our desires so that we do not

---

[1] Aristotle, *Politics*, trans. C. D. C. Reeve (Indianapolis, IN: Hackett, 1998 [c.325 BC]). Hereafter "P."

[2] Leon R. Kass, *The Hungry Soul: Eating and the Perfecting of Our Nature* (Chicago: University of Chicago Press, 1999 [1994]), 92–3, 98.

become the worst of animals in tyrannically seeking to appropriate whatever we want in the world for own use and satisfaction. In other words, as Kass says, what we need first of all is restraint or nay-saying. We need to be able to step back from our desires and ask what it is good to desire, rather than acting on our every whim. But in order to become the best of animals we need not only to restrain our desires; as suggested by the idea of character *formation*, we also need to give shape or *form* to our desires by cultivating dispositions of proper responsiveness toward what is of value in the given world (including our humanity), that is, we need to cultivate the virtues of character.

On Aristotle's account, we become virtuous by repeatedly doing virtuous actions: for example, we become courageous by doing courageous actions, generous by doing generous actions, just by doing just actions, and so on, and the sign that we are acquiring the virtue in each case is that these actions begin to come "naturally" (as part of our second nature) and we have a sense of fulfillment in performing these actions for their own sake as inherently noble and worthwhile activities, and we would feel ashamed if we did not perform these actions in appropriate situations. Of course, we also become vicious by repeatedly doing vicious actions. Thus, in the *Nicomachean Ethics*, Aristotle remarks: "it makes no small difference whether people are habituated in one way or in another way straight from childhood; on the contrary, it makes…*all* the difference" (II.1, 1103b23–5).[3] Earlier he notes: "A nobly brought up person…either has the starting-points or can easily get hold of them" (I.4, 1095b7–8).

The approach to character formation that I want to advance here can be described as both Aristotelian and Confucian in nature, as I want to focus on the central importance of two limiting virtues, namely, moderation, which we find emphasized in the Aristotelian tradition, and reverence, which we find emphasized in the Confucian tradition. I will first begin with the Confucian strand of my approach and then turn to the Aristotelian strand.

At the heart of the Confucian way of life (the *Dao*) is the concept of *li*, which is often translated as "ritual" but where this can be understood to

---

[3] Aristotle, *Nicomachean Ethics*, trans. C. D. C. Reeve (Indianapolis, IN: Hackett, 2014 [*c*.325 BC]). Hereafter "NE."

include traditional norms of propriety, or what we can simply call "good manners."[4] Although *li* has its root meaning in religious ritual (or "sacred ceremony"), Herbert Fingarette points out that Confucius uses "the language and imagery of *li* as a medium within which to talk about the entire body of the *mores*, or more precisely, of the authentic tradition and reasonable conventions of society," which are important for cultivating and expressing virtue.[5] As we saw Kass remark, if we are to become the best of animals, then we need to be perfected "by rearing in customs that bring out and complete what is best in [our] nature." Key to "rearing in customs" here is teaching good manners, that is, patterns of behavior that can serve to habituate one in virtue and give a taste for the noble, and when they become second nature, they can give expression to virtue. In other words, I am suggesting that good manners—e.g., saying "please," "thank you," "you're welcome," "I'm sorry," and "I forgive you," and acting kindly, generously, fairly, considerately, respectfully, and reverently—provide crucial starting points for cultivating and expressing the virtues (e.g., justice, courtesy, generosity, kindness, reverence, and gratitude) and thereby ennobling our humanity. As Fingarette puts it: "Men become truly human as their raw impulse is shaped by *li*. And *li* is the fulfillment of human impulse, the civilized expression of it—not a formalistic dehumanization."[6] In Confucius's *Analects* we find the following teaching: "Restraining yourself and returning to the rites [*li*] constitutes Goodness [*Ren*; this is also translated as Humaneness]" (12.1).[7] Or as the proverbial saying goes: "Manners maketh man."

---

[4] The connection between *li* and manners is also made in Amy Olberding, "Etiquette: A Confucian Contribution to Moral Philosophy," *Ethics* 126:2 (2016): 422–46. Olberding distinguishes between etiquette and manners, saying: "'Etiquette' refers to rules for behavior that conventionally define appropriate conduct within a particular society. In contrast, 'manners' refers to deeper values the rules of etiquette seek to express" (425). One can certainly make this stipulation, but I think of etiquette and manners as basically synonymous. Indeed, it seems strange to think of manners as deeper values; rather, I think they are specific patterns of behavior, which can be good or bad.

[5] Herbert Fingarette, *Confucius: The Secular as Sacred* (Long Grove, IL: Waveland Press, 1972), 6–7.

[6] Fingarette, *Confucius*, 7.

[7] Confucius, *Analects*, trans. Edward Slingerland (Indianapolis, IN: Hackett, 2003 [*c*.450 BC]). The early Confucian philosopher Xunzi writes: "[Ritual] . . . reaches full development in giving [desire] proper form, and finishes in providing it satisfaction. And so when ritual is at its

Indeed, practicing good manners is often seen as part of the civilizing process, where there is an ideal of civility, which contrasts with being barbarous, bestial, crude, etc.[8] This ideal of civility, in fact, captures well Aristotle's idea that it is only in the *polis* (or *civitas* in Latin) with just laws and where virtue is cultivated that we can realize what is noblest in our humanity and thereby become the "best of animals" and achieve our good. The morally—and we might also say spiritually—elevating function of a rearing in custom that involves practicing good manners is expressed well in the following lines from W. B. Yeats's poem "A Prayer for My Daughter":

> And may her bridegroom bring her to a house
> Where all's accustomed, ceremonious;
> For arrogance and hatred are the wares
> Peddled in the thoroughfares.
> How but in custom and in ceremony
> Are innocence and beauty born?

Yeats in fact gives poetic expression not only to the morally elevating function of custom and ceremony (or *li*) but also to the potential for moral anarchy when these are neglected, as seen in the famous lines from "The Second Coming":

most perfect, the requirements of inner dispositions and proper form are both completely fulfilled" (*Xunzi*, ch. 19, trans. Eric L. Hutton, in *Readings in Classical Chinese Philosophy*, ed. Philip J. Ivanhoe and Bryan W. Van Norden [Indianapolis, IN: Hackett, 2001], 266).

[8] See Norbert Elias, *The Civilizing Process, Vol. 1: The History of Manners*, trans. Edmund Jephcott (New York: Pantheon, 1978); Cheshire Calhoun, "The Virtue of Civility," *Philosophy and Public Affairs* 29:3 (2000): 251–75; and Kass, *The Hungry Soul*, 131–2. Roger Scruton writes

> In teaching [children] manners, we are putting the finishing touches on potential members of society, adding the polish that makes them agreeable (Etymologically, "polite" and "polished" are connected...). From the very outset, therefore, we strive to smooth away selfishness. We teach children to be considerate by compelling them to behave in considerate ways. The unruly, bullying, or smart-aleck child is at a great disadvantage in the world, cut off from the lasting sources of human fulfillment. His mother may love him, but others will fear or dislike him.
>
> ("Real Men Have Manners," in *The Philosophy of Food*, ed. David M. Kaplan [Berkeley, CA: University of California Press, 2012], 24–5)

> Things fall apart; the centre cannot hold;
> Mere anarchy is loosed upon the world,
> The blood-dimmed tide is loosed, and everywhere
> The ceremony of innocence is drowned;
> The best lack all conviction, while the worst
> Are full of passionate intensity.

A similar idea is expressed in Cormac McCarthy's *No Country for Old Men*, which is a very Yeatsian novel (indeed, the title is taken from Yeats's poem "Sailing to Byzantium"). Toward the end of the novel, the protagonist, Sheriff Ed Tom Bell, is asked to explain the rise of an extreme and violent form of moral anarchy that he has witnessed in his lifetime and which the reader witnesses in the course of the novel. Ed Tom responds: "It starts when you begin to overlook bad manners. Any time you quit hearin Sir and Mam the end is pretty much in sight."[9]

This might seem like a surprising response. But the thought is, in fact, nothing new. A very similar idea is expressed in Confucius's *Analects*, but the focus is on the effects of good manners rather than those of bad manners: "The gentleman applies himself to the roots. 'Once the roots are firmly established, the Way [*Dao*] will grow.' Might we not say that filial piety [*xiao*] and respect for elders constitute the root of Goodness [*Ren*]?" (1.2). This passage points to what I believe is one of the key functions of good manners, which is that they involve *ways of showing respect or reverence for that which is respect- or reverence-worthy*, which, as we will see, can also reveal this respect- or reverence-worthiness.[10] (Recall that in Chapter 1 I said that reverence can be understood as a heightened form of respect, though I will distinguish between them here). Moreover, this is also important for cultivating and expressing virtues related to showing respect and reverence, such as justice, reverence (or piety), humility, gratitude, and courtesy (or considerateness), which in being properly responsive to that which is respect-worthy or

---

[9] Cormac McCarthy, *No Country for Old Men* (New York: Vintage, 2005), 304.

[10] As Mencius (who is widely regarded as the most important Confucian philosopher after Confucius) puts it: "If one is without the heart of deference, one is not human.... The heart of deference is the sprout of propriety" (*Mencius* 2A: 6, trans. Bryan W. Van Norden, in *Readings in Classical Chinese Philosophy*, 125–6).

reverence-worthy constitute for us a normatively higher, nobler, more fulfilling mode of life.

According to Sarah Buss, a central moral purpose of good manners consists in "appearing respectful" toward others.[11] By "appearing" here she does not mean a mere outward show that is inwardly false. Rather, she believes that the moral requirement of respect for human beings (i.e., acknowledging their equal intrinsic dignity) requires not only that we do not trample on their rights and interests (i.e., use them as a mere means to our ends) but also that we engage in courteous behavior that is expressive of an attitude of respect and avoid rude behavior that expresses disrespect. For instance, saying "please" is a way of respecting someone's dignity and acknowledging his or her autonomy in being able to choose whether to agree to our *request*, which is not the case if we were simply to *demand* something. In this context saying "thank you" also becomes appropriate, because it gratefully acknowledges that someone has freely done a good deed for us.[12] One of Buss's key insights pertains to how courteous behavior is important not only for showing proper respect for others but also for *forming* our sense of human respect-worthiness. Drawing on Cora Diamond's work, she writes:

> [Our] conventions of courtesy influence our assumptions about the moral status of human beings. The countless little rituals we enact to show one another consideration are...the means whereby we "build our notion of human beings." They are "the ways in which we mark what human life is," and, as such, they "belong to the source of moral life."...Good manners...not only inspire good morals. They do so by constructing a conception of human beings as objects of moral concern. To learn that human beings are the sort of animal to whom one must say "please," "thank you," "excuse me," and "good morning," that one ought not to interrupt them when they are speaking, that one ought not to avoid eye contact and yet ought not to stare,...to learn all this and much more is to learn that human beings deserve to be treated

---

[11] Sarah Buss, "Appearing Respectful: The Moral Significance of Manners," *Ethics* 109 (1999): 795–826.

[12] See Buss, "Appearing Respectful," 802; Karen Stohr, *On Manners* (New York: Routledge, 2011), 27–9.

with respect, that they are respectworthy, that is, that they have a dignity not shared by those whom one does not bother to treat with such deference and care.[13]

In order to avoid the moral constructivist or non-realist impression that might be given by these remarks about "constructing a conception of human beings as objects of moral concern"—which, if endorsed, would undermine the moral phenomenology of respect-*worthiness* at issue—we should say that these manners (or "little rituals") and the conception of human beings they help to "construct" or "build up" enable us to *see* and *experience* what human beings *really are*, namely, they are respect-worthy, and thus that they are *fittingly* treated and regarded in these ways. In other words, these manners are *revelatory*; they enable a *transfigured* or *regestalted* vision whereby the equal inherent respect-worthiness of all human beings can come into view. This is a revelation that depends upon the *enactment* of this respect-worthiness through manners in that we only come to grasp fully the significance at issue through living it out.[14]

Buss's remarks in the cited passage are the most direct comments she makes about the importance of manners for character formation, though she does not explore this in much detail. These remarks, in fact, connect up with and lend support for the truth of Ed Tom Bell's remarks in *No Country for Old Men*: "Any time you quit hearin Sir and Mam the end is pretty much in sight." The idea here is that if young people do not practice courteous behavior, then they will fail to build up a conception of human beings as inherently worthy of equal respect such that they set limits upon one's will. Failing this, "anything goes." Or it also might be that bad manners do not lead to complete moral anarchy but to bad

---

[13] Buss, "Appearing Respectful," 800–1. Cora Diamond writes:

> [It] is not out of respect for the interests of beings of the class to which we belong that we give names to each other, or that we treat human sexuality or birth or death as we do, marking them—in their various ways—as significant or serious. And again, it is not respect for our interests which is involved in our not eating each other. These are all things that go to determine what sort of concept "human being" is.
>
> ("Eating Meat and Eating People," *Philosophy* 53, no. 206 [1978]: 469–70)

[14] On enacted meaning, see Charles Taylor, *The Language Animal: The Full Shape of the Human Linguistic Capacity* (Cambridge, MA: Belknap Press of Harvard University, 2016), 42–7.

morality. For example, we might think of a community with racist norms in which only one's own race is treated with deference and care, while other races are treated with disrespect. Instead of building up a conception of human beings as inherently worthy of equal respect, here one builds up a conception of human beings that *blinds* one from seeing and experiencing the equal inherent dignity of humanity. We can say something similar about certain classist attitudes, where manners are used to distinguish one's social class from others' and to denigrate those who are not of one's class. Thus, a lot depends upon being brought up with *good* manners (i.e., those that enable us to see and appropriately respond to moral reality), rather than bad ones (i.e., those that blind us to moral reality); and if we happen to be brought up in manners that are to some degree bad, then it is important that we can reform these bad manners and acquire good ones.

While Buss's account in the cited passage makes some crucial points, I think we need to go beyond it and see that manners can express not only respect (for what has dignity) but also reverence (for what is sacred or holy). Moreover, the sense of respect or reverence that manners can express is directed not only toward the equal inherent dignity or sanctity of human life; it can also be directed toward the specific achievements or good deeds or roles of others. Further, it is not always directed toward human beings. We can see how all of this is so by looking in more detail at the role of teaching good manners in the process of character formation.

The first context in which children learn to show respect and reverence is of course the home. Although saying "Sir" and "Ma'am" may not be necessary with regard to one's parents—indeed it is perhaps inappropriate in so far as it conveys an emotional distance—it is still important that children learn appropriate ways of honoring their mother and father and thereby acquire the virtue of filial piety.[15] The respect or reverence owed to parents here does not primarily have to do with their intrinsic

---

[15] For an illuminating discussion of the biblical commandment to honor one's mother and father in the Ten Commandments, see Marilynne Robinson, *Gilead* (New York: Farrar, Straus and Giroux, 2004), 134–9. At the end the protagonist remarks:

> There's a pattern in these Commandments of setting things apart so that their holiness will be perceived. Every day is holy, but the Sabbath is set apart so that the holiness of time can be experienced. Every human being is worthy of honor, but the conspicuous discipline of honor is learned from this setting apart of the mother

dignity: it is owed in virtue of what they have given and will give. First of all, parents give the gift of life (which is often seen as something sacred or profoundly precious). The "gift" here does not have to be intentional, though it may be. The point is that one's parents are sources of one's existence, and if such existence is regarded as having great value (and hence as a gift), then filial piety and gratitude are appropriate (something similar could be said about other sources of one's existence). Secondly, parents—when acting as they should—give life-sustaining care and provide education that enables a child to move toward realizing what is most admirable in the child's human potential and thereby achieving his or her good. Here filial piety and gratitude are again appropriate. In this context filial piety or respect for parents is also important because it enables a needed kind of humility, namely, "docility," understood as being *teachable*, including with regard to good manners and the virtues. Here we see a key aspect of the truth in the rhetorical question of the *Analects* cited earlier: "Might we not say that filial piety [*xiao*] and respect for elders constitute the root of Goodness [*Ren*]?" Children need to begin by trusting their parents to lead them in the way (*Dao*) of truth and goodness. It is possible that this trust will turn out to be misplaced, as unfortunately there are bad parents. But this is something a child does not fully know until later on, and he or she will still need parent-surrogates who can provide the sort of education required for realizing his or her good. In short, filial piety and trust of parents or parent-surrogates are necessary for acquiring and exercising all the virtues.

As part of teaching virtue, good parents will teach their children manners that express respect or reverence not only for all human beings in virtue of their intrinsic dignity but also for teachers, elders, ancestors, traditions, customs, institutions, community, sexuality, life, death, etc. It is in regard to teachers and elders where saying "Sir" or "Ma'am," "Mr." or "Ms.," "Dr.," and other honorific titles are particularly appropriate.

and father, who usually labor and are heavy-laden, and may be cranky or stingy or ignorant or overbearing. Believe me, I know this can be a hard Commandment to keep. But I believe also that the rewards of obedience are great, because the root of real honor is always the sense of the sacredness of the person who is its object. (139)

This can communicate respect for teachers and elders in virtue of the good things they have to offer as well as openness to instruction. Filial piety can, in fact, be understood in an expanded sense to include respect or reverence not only for parents but also for teachers, elders, ancestors, traditions, and the community and its laws and customs. Each of these can be seen to have a parent-like role in guiding us toward the realization of our good, and if so, respect or reverence is owed to them along with an appropriate openness to instruction (which is compatible with criticism, especially as one matures).

In the case of elders, there is also a reverence owed to human life itself as embodied in the lifespan, and something similar can be said for ancestors. The proper reverence for human life is, in fact, expressed in many forms of good manners. For instance, certain reverent behavior is expected at funerals. Something is similarly true regarding rituals welcoming and honoring new life, such as baptism, and likewise with regard to traditional wedding ceremonies, though here reverence is due to the profound nature of human sexuality and the bond of romantic love ("holy matrimony"), as well as to the possibility of new life resulting from it. In general, *sacred things*—i.e., that which is reverence-worthy and "set apart"—require certain *reverent manners*, and *dignified things* require certain *respectful manners*.

The rites of passage mentioned in the preceding paragraph recall Buss's remarks about the "countless little rituals" whereby we build up our conception of human respect-worthiness. But, as we have seen, these rituals are not necessarily "little" and they can express respect or reverence not only for particular human beings but also for life, death, sexuality, tradition, custom, etc. Good manners or *li*, I am also suggesting, go to build up a sense of human beings and other features of our world as respect-worthy or reverence-worthy. Indeed, Confucius's use of the language and imagery of religious ritual to understand traditional manners is intended to enable us to see everyday life as imbued with profound dignity and sanctity; as Fingarette puts it, "The image of the Holy Rite as a metaphor of human existence brings foremost to our attention the dimension of the holy in man's existence."[16]

---

[16] Fingarette, *Confucius: The Secular as Sacred*, 16.

A main lesson to be drawn from the foregoing is that if we throw off all customary manners or rituals, then we do so at our moral peril. The person who discards all social conventions or rituals does not, in fact, see most clearly but instead becomes blind to the reality of that which is respect-worthy and reverence-worthy and so invites moral nihilism.[17] The "countless little [and big] rituals" in human life are not mere add-ons to the way the world is; rather, they provide the lenses through which moral reality can come into view, namely, that which is respect-worthy or reverence-worthy and which places restraints on our will.

Let us now turn to the Aristotelian strand of my approach to character formation and focus on the importance of the limiting virtue of moderation. Moderation is sometimes seen as equivalent to temperance (indeed the verbs "to moderate" and "to temper" are essentially synonymous), but I want to treat moderation as the broader concept and temperance as an instance of moderation concerned with sensual

---

[17] This point is depicted well by Flannery O'Connor. In her novel *The Violent Bear It Away* (published in 1960), the 14-year-old protagonist Francis Tarwater begins a path of nihilistic destruction and violence by neglecting to perform a proper Christian burial for his great uncle Mason, and a baptism for his cognitively disabled cousin Bishop. This shows not only a failure of reverence for life and death but also a lack of filial piety, since the performance of both rituals was requested by Mason, who raised and cared for him. Regarding the burial, Francis says: "The dead are poor...You can't be poorer than dead. He'll have to take what he gets.... Now I can do anything I want to...Could kill off all those chickens if I had a mind to" (*Collected Works* [New York: Library of America, 1988], 345). Later he says: "You don't owe the dead anything" (363). By contrast, Mason says: "The world was made for the dead" (339). As for baptism, it is described as a "gesture of human dignity" (351). It is significant then that Mason regarded it as so important that Bishop be baptized, suggesting that he possesses the same human dignity as anyone else. But Francis regards Bishop as follows:

> He's like a hog...He eats like a hog and he don't think no more than a hog and when he dies, he'll rot like a hog. Me and you too...will rot like hogs. The only difference between me and you and a hog is me and you can calculate, but there ain't any difference between him and one.   (403)

Bishop's own father, the scientist utilitarian Rayber, remarks: "Nothing ever happens to that kind of child...In a hundred years people may have learned enough to put them to sleep when they're born." Francis's emotional response to these remarks is a "war between agreement and outrage" (435). Later a woman says to Francis: "Mind how you talk to one of them there, you boy!" He responds: "Them there what?" And the woman says: "That there kind," as she "[looks] at him fiercely as if he had profaned the holy" (426–7). In much of her work O'Connor is concerned with nihilism and its connection with bad manners. For instance, in her short story "Good Country People" (published in 1955), bad manners (namely, rudeness) in the main character, Hulga, are indicative of a nihilistic perspective (Hulga says: "I don't have illusions. I'm one of those people who see through to nothing" [280]) and portend ruin. In a letter O'Connor writes: "Somewhere is better than anywhere. And traditional manners, however unbalanced, are better than no manners at all" (856).

pleasure, which is how Aristotle treats temperance (*sōphrosunē*). The virtue of moderation is concerned with avoiding vicious extremes, and hence it can be regarded as a limiting virtue. Moderation can, in fact, be regarded as the master virtue of character for Aristotle, since he describes a virtue of character as a "medial condition" between the extremes of excess and deficiency with regard to some feeling or action, where the proper mean is determined through practical wisdom (NE II.6, 1106b35–1107a5). For instance, the virtue of courage concerns the feeling of fear and is a mean between cowardice (the vice of excessive fear) and rashness (the vice of deficient fear). Likewise, the virtue of generosity concerns the action of giving and is a mean between waste-fulness (the vice of excessive giving) and stinginess (the vice of deficient giving).

Here I do want to focus on moderation in the form of temperance, because it is of central significance for the process of character formation (in Chapter 3 I will discuss the importance of political moderation for avoiding and counteracting political polarization). Aristotle notes that the young typically "live in accord with their feelings and pursue most of all what is pleasant for themselves and present at hand" (NE VIII.3, 1156a30–3), and this is why he thinks that they are often not proper students for his lectures on ethics:

> Argument and teaching . . . surely do not have the strength in everyone but, rather, the soul of the audience must be prepared beforehand through habits to enjoy and hate in a noble way, like earth that is to nourish seed. For someone who lives in accord with his feelings will not listen to—or, what is more, comprehend—argument that encourages him to turn away. And in a state like that how is it possible to persuade him to change his ways? . . . Character, then, must in some way be there beforehand and properly suited for virtue, liking what is noble and repelled by what is shameful.   (NE X.9, 1179b22–30)

As we saw earlier, Aristotle thinks it makes all the difference whether or not we are habituated in the practice of the virtues (temperance, courage, generosity, justice, etc.) from childhood and thereby acquire a taste for what is noble and a repulsion from what is shameful.

Before considering Aristotle's account of temperance, it is worth noting that he thinks proneness to feeling shame can be a germ of virtue in young people. He remarks:

> [We] think that young people should have a sense of shame because they live by their feelings and so make many errors but are restrained by shame.... No one would praise an older person for being prone to shame, however, since we think that he shouldn't do any actions that give rise to shame.   (NE IV.9, 1128b16–20)

The awakening of a sense of shame is, in fact, key to our coming into ethical self-consciousness and therefore for our humanization—i.e., for realizing what is most admirable in our humanity—since non-human animals are incapable of the feeling of shame.[18]

Temperance is also key to our humanization, precisely by helping us to avoid being bestial and enabling us to pursue what is noble both in the realm of sensual pleasure and beyond. Aristotle says that temperance (*sōphrosunē*) is the mean between intemperance and insensibility (NE II.7, 1108a4–8). The latter he says rarely occurs, "since that sort of insensibility is not human" (III.11, 1119a5–6). So his focus is on avoiding intemperance, or *akolasia*, which can also be translated as

---

[18] This is depicted well in the story of the Garden of Eden in Genesis 2:4–3:24, where Adam and Eve achieve ethical self-consciousness through their transgression in eating the forbidden fruit from the "tree of the knowledge of good and evil." The story can be understood as symbolizing the humanization that occurs in each of our lives as we emerge from childhood innocence into ethical self-consciousness, often through the experience of shame resulting from transgression. While such transgression is regrettable and should never be sought, in so far as it is bound up with the emergence of ethical self-consciousness and therefore is part of our humanization, we can also regard it as a "happy fault": it is a condition for the possibility of realizing a higher mode of life than is possible for non-rational animals. One might wonder whether the story of the Garden of Eden is meant to suggest that it would be better for humans to remain at the level of non-human animals: after all, God does forbid eating from the tree of the knowledge of good and evil. But if we acknowledge that God thought it "very good" to create human beings in the "image of God" (as Genesis 1 affirms), then we should instead read the story as teaching us about the perils of rational, ethical self-consciousness for beings who are not God but instead are situated between the beasts and the divine, where what is most perilous is the hubristic tendency to "play God." For more on these matters, see Leon R. Kass, *The Beginning of Wisdom: Reading Genesis* (Chicago: University of Chicago Press, 2003), chs. 2 and 3.

"unrestrainedness" with regard to our animal appetites.[19] Aristotle remarks: "Temperance and intemperance are concerned with the sorts of pleasures that the rest of the animals share in as well (which is why they appear slavish and beast-like), namely, touch and taste.... To enjoy such things,... and to like them most, is beast-like" (III.10, 1118a23–5). In order to avoid becoming bestial such animal appetites need to become obedient to reason, which is what makes for temperance. Aristotle says of the temperate person: "pleasant things that are conducive to health or a good state, these he will desire moderately and in the way he should, as he will the other pleasant things that do not impede these or are not contrary to what is noble or beyond his means" (III.11, 1119a15–18). Aristotle goes on to remark: "[If] the appetitive element is not obedient and subordinate to the ruling element [i.e., reason], it will grow and grow. For the desire for pleasure is insatiable...and if the appetites are large and intense, they even knock out rational calculation" (III.12, 1119b6–9). We have a choice here between ruling our animal appetites through reason or being ruled by them.[20] Aristotle remarks: "[A] temperate person's appetitive element should be in harmony with reason. For the target of both is what is noble, and a temperate person has an appetite for the things he should and in the way and when he should, which is just what reason, for its part, prescribes" (III.12, 1119b14–17).

In his book *The Hungry Soul*—which I cited at the outset of this section—Leon Kass writes: "Moderation keeps us free from enslavement to desire and free for receptivity and responsiveness," namely, to that which is noble or makes for nobility.[21] He goes on to say that it may be

---

[19] See Kass, *The Hungry Soul*, 155. In his glossary entry on "intemperance, *akolasia*," Terence Irwin writes: "This term for the vice of excess—opposed to temperance—is derived from *kolazein* ('punish, correct'), and so indicates someone whose desires lack the corrective treatment they need to make them subject to correct reason" (Aristotle, *Nicomachean Ethics*, 2nd edn, trans. Terence Irwin [Indianapolis, IN: Hackett, 1999], 336).

[20] See Kass, *The Hungry Soul*, 156.

[21] Robert C. Roberts holds a similar view:

Temperance is that state of character in which bodily appetites successfully conform to the larger concerns of the moral life.... Seldom if ever do adult human beings have purely "physical" appetites and pleasures; these virtually always have a human "meaning" of one sort or another that affects the feeling of the appetite and the pleasure of its satisfaction. The virtue of temperance is thus a disposition to have

tempting to think that such nobility "consists in 'rising above' or 'transcending' our animality," but he counters: "'Nobility' is not so much a transcendence of animality as it is the turning of animality into its peculiarly human and regulated form" (158–9). Kass's own concern here is with the realm of eating, and he seeks to show how we can eat in a way that contributes to the "perfecting of our nature." He discusses how certain customs help to tame our human omnivorousness (which is expressive of our "broad, open, and undetermined appetites") and transform animal feeding into distinctively human eating, where we realize what is most admirable in our humanity in this context. For instance, we eat around a table (or a similar surface) with others in order to foster conversation, family life, friendship, and refinement. Moreover, there are basic *table manners*—e.g., sitting upright; using utensils; being neat; acting temperately in not overeating or eating too fast or otherwise being enslaved to animal appetite; avoiding rude behavior such as burping, chewing with one's mouth open, or talking with food in one's mouth—and their purpose is to foster human communion around a meal, avoid putting off others, and ennoble, dignify, and beautify our animal necessity and give expression to virtues such as temperance, generosity, and gratitude. Such "civilized" eating enables the possibility of higher modes of human experience found in fine dining and feasting. While both are ennobled forms of eating, feasting also involves the "sanctification of eating," where saying a prayer or blessing over the meal is especially important (though it is arguably important in all human eating). Here, in an activity rooted in our animal necessity, we reach the heights of human self-consciousness, as feasting involves a celebration of the world and our place within it.[22] Kass writes: "one can speak also about piety and reverence, and the human impulse toward

> appetites and pleasures that are qualified by *appropriate* concerns and *not by inappropriate* concerns, and thus are *appropriate* desires, and *appropriate* pleasures.
>
> ("Temperance," in *Virtues and Their Vices*, ed. Kevin Timpe and Craig Boyd [Oxford: Oxford University Press, 2014], 103–4)

This relates to the discussion that follows on animal feeding versus human eating and lust versus romantic love.

[22] Consider Josef Pieper's account of "festivity" (where the traditional religious feast is the key reference point):

> Underlying all festive joy...there has to be an absolutely universal affirmation extending to the world as a whole, to the reality of things and the existence of

transcendence, beginning in awe and fear and, sometimes encouraged by wine, moving through feelings of gratitude and songs of praise in the direction of encountering the divine" (163; Kass discusses Isak Dinesen's short story "Babette's Feast," where this is well-depicted). In short, when "the hungry soul" comes to the table, he or she desires not only bodily nourishment but also spiritual nourishment: the hungry soul yearns for beauty, nobility, conviviality, and even sanctity.

If we consider sexual desire, we see how a similar transformation can occur through "rearing in customs that bring out and complete what is best in [our] nature." Like human omnivorousness, human sexual desire can also be seen as representative of our troublesome "broad, open, and undetermined appetites." In order to ennoble and thereby humanize sexual desire, we must transform mere *lust* (which is common in animal life) into *romantic love* (which is distinctively human), which can be, at its best, one of the highest modes of human experience, and where the categories of the noble and the sacred also find expression. And this again will require good manners that cultivate and express virtues in the sexual domain, such as chastity (that is, right feeling and intention in sexual matters), fidelity (which is traditionally connected to marital vows), modesty, and romantic love itself. To recall Yeats's poem "A Prayer for My Daughter": "How but in custom and in ceremony are innocence and beauty born?"

Absolute Prohibitions and the Virtue of Reverence

In the preceding section I said that the limiting virtue of moderation can be regarded as the master virtue of character for Aristotle, given his account of a virtue of character as a "medial condition" between the extremes of excess and deficiency with regard to some feeling or action.

---

man himself.... Festivity, in its essential core, is nothing but the living out of this affirmation. *To celebrate a festival means: to live out, for some special occasion and in an uncommon manner, the universal assent to the world as a whole.*
(*In Tune with the World: A Theory of Festivity*, trans. Richard and Clare Winston [South Bend, IN: St. Augustine's Press, 1999 (1963)], 26, 30)

To this account of the virtues of character, Aristotle adds the following qualifier:

> But not every action or every feeling admits of the medial condition, since in some cases they are named in such a way that they are united with baseness from the start—for example, spite, shamelessness, and envy, and (in the case of actions) adultery, theft, and murder. For all these and things like them—and not the excessive varieties of them or the deficient ones—are said to be what they are because they are base. It is never possible, then, to be correct where they are concerned but it is necessary always to be in error.   (NE II.6, 1107a8–14)

This passage suggests that Aristotle accepts absolute prohibitions, which puts him at odds with consequentialist ethics. Indeed, later he writes: "In some cases ... there is presumably no being compelled. On the contrary, rather than do [the actions] we should die having suffered the most terrible things" (III.1, 1110a26–8). He mentions the act of killing one's own mother as such a case where it would be better instead to "die having suffered the most terrible things."

Aristotle leaves a lot unexplained in both passages. For instance, what explains the baseness of murder such that it always ought to be avoided? The evil of murder, in my view, needs to be explained in terms of how it gravely violates the intrinsic value of human life, which should be regarded as sacred or reverence-worthy. This is suggestive of the general approach I want to take here: I want to argue that in order to justify and make sense of absolute prohibitions of certain actions—I set aside Aristotle's remarks on feelings that are absolutely prohibited—we need a sense of the sacred or the reverence-worthy, which means that we need the limiting virtue of reverence, which recognizes absolute limits on our will as part of what is involved in being properly responsive to that which is sacred or reverence-worthy. This approach is suggested, I think, by Aristotle's example of the grave evil of killing one's own mother, which seems especially irreverent or impious, since one's mother is a source of one's life.

I will make the case for a reverence-based approach to absolute prohibitions by contrasting it with other prominent approaches to defending absolute prohibitions that I will argue are unsuccessful. These other approaches are the following: (1) the preventing moral anarchy approach; (2) the moral identity approach; (3) the character-centered approach; and (4) the divine law approach. Let us consider each in turn.

Stuart Hampshire is a good example of someone who adopts the *preventing moral anarchy approach*. He is concerned to argue against utilitarianism as a prominent form of consequentialist ethics that rejects absolute prohibitions because of maintaining that the ends justify the means. In particular, utilitarianism maintains that we should seek to promote the greatest happiness for the greatest number, where happiness is understood in terms of pleasure and the absence of pain, or in terms of preference-satisfaction. Hampshire writes:

> [There] is one feature of familiar moralities which utilitarian ethics famously repudiates, or at least makes little of. There are a number of different moral prohibitions, apparent barriers to action, which a man acknowledges and which he thinks of as more or less insurmountable...For example, in addition to certain fairly specific types of killing, certain fairly specific types of sexual promiscuity, certain takings of property, there are also types of disloyalty and of cowardice, particularly disloyalty to friends, which are very generally, almost universally, forbidden and forbidden absolutely. They are forbidden as being intrinsically disgraceful and unworthy, and as being, just for these reasons, ruled out: ruled out because they would be disgusting, or disgraceful, or shameful, or brutal, or inhuman, or base, or an outrage. In arguing against utilitarians I must dwell a little on these epithets usually associated with morally impossible action, on a sense of disgrace, of outrage, of horror, of baseness, of brutality, and, most important, a sense that a barrier, assumed to be firm and almost insurmountable, has been knocked over, and a feeling that, if this horrible, or outrageous, or squalid, or brutal action is possible, then anything is possible and nothing is forbidden, and all restraints are threatened.[23]

---

[23] Stuart Hampshire, *Morality and Conflict* (Cambridge, MA: Harvard University Press, 1983), 87–9.

Hampshire acknowledges that "these ideas have often been associated with impiety," where the concern with impiety is understood in terms of a fear of inciting divine anger. However, he maintains that such ideas do not need to have this association. Instead, he thinks a purely secular account can be given in terms of "a fear of human nature," where there is always the possibility of falling into a Hobbesian state of nature, where there is a war of all against all and in which life is poor, nasty, brutish, and short. The outrage that Hampshire describes is, he thinks, "an aspect of respect for morality itself rather than for any particular morality and for any particular set of prohibitions" (89). In other words, the idea here is we need *some* form of morality that prevents us from falling into moral anarchy, where everything is permissible.

What should we make of this? The biggest problem with this approach is that it is using what are, in fact, consequentialist considerations about falling into moral anarchy in order to argue against consequentialist forms of ethics, and in doing so it cannot rule out that there are some cases in which the prohibited action can be done without it leading to moral anarchy. I do not deny that many people may fear moral anarchy if certain horrible actions are not absolutely prohibited, but I do not think these prohibitions should be seen as ultimately grounded in a "respect for morality itself rather than for any particular morality and for any particular set of prohibitions." It is not clear what "morality itself" is supposed to be when this is understood apart from any particular conception of morality; it certainly cannot be a utilitarian conception of morality. We need to inhabit a particular conception of morality with a particular set of prohibitions, since the epithets associated with morally impossible actions to which Hampshire appeals—e.g., outrageous, horrible, brutal, and inhuman—depend upon there being features of the world that make them appropriate and which demand certain responses, and they cannot just be based on the thought that without certain moral beliefs we will fall into moral anarchy. Consider again the case of murder. What is horrible about murder is not simply that if we allow it, then we will fall into a state of moral anarchy (though this may be true, at least if it becomes widespread enough); rather, it is horrible and wrong because it gravely violates the special dignity or sanctity of human life. Hampshire is right to note the traditional association between impiety (or irreverence) and morally impossible actions, but he is wrong to understand the concern with impiety in terms of a fear of inciting divine anger. The central concern of piety (or reverence) is proper responsiveness to the reverence-worthy.

Hampshire also offers another approach to defending absolute prohibitions, which I call the *moral identity approach*. He writes:

> I believe that critical reflection may leave the notion of absolutely forbidden, because absolutely repugnant, conduct untouched.... [There] may be reflective reasons, in the sense that one is able to say why the conduct is impossible as destroying the ideal of a way of life that one aspires to and respects, as being, for example, utterly unjust or cruel or treacherous or corruptly dishonest.   (90)

The idea here is that avoiding certain actions is a necessary condition for maintaining one's moral identity. But is Hampshire right that critical reflection will leave the notion of the absolutely forbidden untouched if one can connect it with "a way of life that one aspires to and respects"? Not necessarily, since one thing we may ask on critical reflection is the following: why should I have *this* particular moral identity? Here we can confront what Bernard Williams identifies as the problem of radical contingency: on critical reflection we may come to see our moral beliefs and moral identity as entirely dependent on the contingencies of our personal, cultural, and evolutionary histories. Williams writes: "This sense of contingency can seem to be in tension with something that our ethical ideas themselves demand, a recognition of their authority."[24] In order to overcome this problem, we need to see our moral identity as being responsive to objective moral demands that are regarded as being there whether or not we are responsive to them. And in the case of affirming absolute prohibitions, I am suggesting that these moral

---

[24] Bernard Williams, *Truth and Truthfulness* (Princeton, NJ: Princeton University Press, 2002), 21. In *Ethics and the Limits of Philosophy* (Cambridge, MA: Harvard University Press, 1985), Williams similarly discusses how reflection can undermine what we take to be ethical knowledge, especially as we "become conscious of ethical variation and of the kinds of explanation it may receive" (159; see also 142–8, 158–9, 163–4, 167–70, 199–200).

demands need to be connected with the sacred or reverence-worthy, to which the virtue of reverence seeks to be properly responsive.

One might think that perhaps we can get around appeals to the sacred or the reverence-worthy if we adopt a *character-centered approach* that focuses on the types of character traits (i.e., the virtues) that are needed in order to flourish as a human being, and where certain of these character traits are seen as requiring us to acknowledge absolute prohibitions. Note that I also appeal to a virtue, namely the virtue of reverence, but the focus of concern is on that which is reverence-worthy or sacred, which is seen as making moral demands upon us. The character-centered approach that I am discussing here is agent-centered. Consider, for instance, Rosalind Hursthouse's remarks on the wrongness of killing: "What is wrong with killing, when it is wrong, may not so much be that it is unjust, violating the right to life, but, frequently, that it is callous and contrary to the virtue of charity."[25] I think this agent-centered explanation gets things backwards: someone is callous in this case *because* he or she fails to be properly responsive to the value of human life, that is, to its special dignity or reverence-worthiness.

It is also worth considering what Elizabeth Anscombe has to say on these matters, since in her influential essay "Modern Moral Philosophy" she raised the issue of whether a virtue ethics approach focused on promoting one's flourishing as a human being can be used to justify absolute prohibitions, and she suggested the following possibility:

One man…may say that since justice is a virtue, and injustice a vice, and virtues and vices are built up by the performances of the action in which they are instanced, an act of injustice will tend to make a man bad; and essentially the flourishing of a man *qua* man consists in his being good (e.g. in virtues); but for any X to which such terms apply, X needs what makes it flourish, so a man needs, or ought to perform, only virtuous actions; and even if, as it must be admitted may happen, he flourishes less, or not at all, in inessentials, by avoiding injustice, his

---

[25] Rosalind Hursthouse, *On Virtue Ethics* (Oxford: Oxford University Press, 1999), 6; see also 57–8, 72, 83–7.

life is spoiled in essentials by not avoiding injustice—so he still needs to perform only just actions.[26]

However, Anscombe expresses skepticism about this sort of argument:

[It] is a bit much to swallow that a man in pain and hunger and poor and friendless is flourishing, as Aristotle himself admitted. Further, someone might say that one at least needed to stay alive to flourish. Another man unimpressed by all that will say in a hard case "What we need is such-and-such, which we won't get without doing this (which is unjust)—so this is what we ought to do."[27]

Paralleling a common critique of rule-utilitarianism, it seems that for the sake of the goal of realizing our own flourishing as human beings (at least when this is understood in a non-moralized way) we could always make exceptions to the demands of virtue.

In "Modern Moral Philosophy," Anscombe ultimately seems to endorse a *divine law approach* to validating absolute prohibitions. She writes:

The man who believes in divine laws will say perhaps "It is forbidden, and however it looks, it cannot be to anyone's profit to commit injustice"; he like the Greek philosophers can think in terms of flourishing.... [If] he is a Jew or Christian, he need not have any

---

[26] G. E. M. Anscombe, "Modern Moral Philosophy," in *Ethics, Religion and Politics: Collected Philosophical Papers, Volume III* (Minneapolis, MN: University of Minnesota Press, 1981), 41.

[27] Anscombe, "Modern Moral Philosophy," 41–2. Consider also Peter Geach:

[Somebody] might very well admit that not only is there something bad about certain acts, but also it is desirable to become the sort of person who needs to act in the contrary way; and yet not admit that such acts are to be avoided in all circumstances and at any price. To be sure, a virtuous person cannot be ready in advance to do such acts; and if he does do them they will damage his virtuous habits and perhaps irreparably wreck his hard-won integrity of soul. But at this point someone may protest "Are you the only person to be considered? Suppose the price of your precious integrity is a most fearful disaster! Haven't you got a hand to burn for your country (or mankind) and your friends?" This sort of appeal has not, I think, been adequately answered on Aristotelian lines.

(*God and the Soul* [London: Routledge, 1978], 123)

very distinct notion: the way it will profit him to abstain from injustice is something that he leaves to God to determine, himself only saying "It can't do me any good to go against his law." (He also hopes for a great reward in a new life later on, e.g. at the coming of Messiah; but in this he is relying on special promises.)   (42)

The problem with this approach is that while it does provide the religious believer with some extrinsic reasons for living in accordance with absolute prohibitions that have been divinely decreed, it does not account for what should be seen as the intrinsic reasons for absolute prohibitions, namely, they concern that which is sacred or reverence-worthy and thus should be regarded as inviolable. For instance, the reason why we should never intentionally kill an innocent human being is because doing so violates the special dignity or sanctity of human life. Unless a theist embraces theological voluntarism (where something is thought to be right or wrong simply *because* God willed it to be so, and which is a position that should be rejected because it makes morality arbitrary and undermines God's praiseworthiness), he or she will believe that God decreed certain absolute prohibitions *because* they concern that which is sacred or reverence-worthy and thus should be regarded as inviolable.

In her later work Anscombe in fact invokes something like reverence when she appeals to what she calls "mystical perception" in order to make sense of certain moral demands upon us, including absolute prohibitions. What such perception seems essentially to involve is a recognition of that which is sacred or reverence-worthy. Anscombe thinks that this perception is "as common as humanity"; for example, it is present in the perception that we dishonor our bodies in casual sex, in our sense that we owe respect to someone's dead body, and in our horror at the evil of murder.[28] In light of this, she distinguishes between two kinds of virtue. Some virtues, such as temperance in regard to food and drink and honesty about property, "are fundamentally utilitarian in character." "Utilitarian" here just means that they are instrumental to

---

[28] G. E. M. Anscombe, "Contraception and Chastity," in *Faith in a Hard Ground: Essays on Religion, Philosophy and Ethics by G. E. M. Anscombe*, ed. Mary Geach and Luke Gormally (Charlottesville, VA: Imprint Academic, 2008), 186–7.

things going well for us. By contrast, some virtues, "though indeed profitable, are supra-utilitarian and hence mystical." One example of a mystical or supra-utilitarian virtue is what Anscombe describes as the virtue of "respect for life," which can also be understood as the virtue of reverence. Although the prohibition on murder certainly "makes life more commodious," she says:

> everybody perceives quite clearly that the wrong done in murder is done first and foremost to the victim, whose life is not inconvenienced, it just isn't there any more. He isn't there to complain: so the utilitarian argument has to be on behalf of the rest of us. Therefore, though true, it is highly comic and is not the foundation: the objection to murder is supra-utilitarian.   (187)

Elsewhere Anscombe speaks similarly of a "religious attitude" of "respect before the mystery of human life"—or what I would call the sacredness or reverence-worthiness of human life—which is "not necessarily connected only with some one particular religious system."[29] She writes:

> A religious attitude may be merely incipient, prompting a certain fear before the idea of ever destroying a human life, and refusing to make a "quality of life" judgment to terminate a human being. Or it may be more developed, perceiving that men are made by God in God's likeness, to know and love God.   (270)[30]

In my discussion here I have focused on murder (i.e., the intentional taking of innocent human life) as a paradigm case of something absolutely prohibited, and I think Anscombe's remarks help to support my contention that we need a sense of the sacred or the reverence-worthy in order to make sense of the wrong of murder and why it should be absolutely prohibited (despite the good that can sometimes come from

---

[29] G. E. M. Anscombe, "Murder and the Morality of Euthanasia," in *Human Life, Action and Ethics: Essays by G. E. M. Anscombe*, ed. Mary Geach and Luke Gormally (Charlottesville, VA: Imprint Academic, 2005), 269–70.

[30] I have discussed Anscombe's views at greater length in *Virtue and Meaning*, ch. 3. There is some overlap in the discussion here.

it, such as in dropping a bomb on innocent civilians to end a war). The same sort of case can be made for any other action legitimately considered to be absolutely prohibited. For instance, adultery is absolutely prohibited because of the sacred trust that is broken (which was established through the solemn vow of marriage), which also constitutes a disrespect for one's spouse and for the profound intimacy shared in spousal sexual union. Similarly, if we think lying is absolutely prohibited, then it should be because truth is sacred and/or the trust between people is sacred. The main point here is that we need to cultivate the virtue of reverence if we are to make sense of absolute prohibitions.

Duties of Assistance: Neighborliness and Loyalty as Limiting Virtues

We have been considering absolute prohibitions as fundamental moral limits, but in this section I want to examine moral limits in another sense, which concern our positive duties of assistance, rather than our negative duties of avoiding wrongful acts (as with absolute prohibitions). Here we are considering the limits on morality itself with respect to what we can be required to do on behalf of others. Just as with my discussion of absolute prohibitions, I will take utilitarianism to provide the main contrast case to the position I will be defending.

One striking difference between utilitarianism and prominent ancient views of ethics is in the place of partiality in the ethical life.[31] We have already seen that Confucian ethics places a strong emphasis on filial piety and the importance of family relationships. As Mencius puts it: "The Way lies in what is near, but people seek it in what is distant...If everyone would treat their kin as kin, and their elders as elders, the world would be at peace."[32] It is also noteworthy that in the *Nicomachean Ethics*, Aristotle spends more time on the topic of friendship (*philia*) than on any other topic. By contrast, utilitarianism endorses strict impartiality in putting forward the moral requirement of promoting the greatest

---

[31] The same is true for Kantianism, though my focus here is on utilitarianism.
[32] *Mencius* 4A11, in *Readings in Classical Chinese Philosophy*, 134.

happiness for the greatest number. John Stuart Mill writes: "As between his own happiness and that of others, utilitarianism requires [one] to be as strictly impartial as a disinterested and benevolent spectator."[33] The implication of this impartiality requirement is not only that self-preference is morally unjustified, but so is partiality to family over non-family, friends over non-friends, and fellows citizens over citizens of other countries.[34] Some utilitarians have tried to make the case for special concern for family, friends, and fellow citizens as the best way to promote the general happiness, but such an indirect form of justification for partial concern does not capture the reasons why most have such concern (as Bernard Williams put it, these indirect forms of justification for particular concern involve "one thought too many"[35]), and it seems that whenever we could show that such partial concern is not justified by the utilitarian standard, it would have to be forgone.

The utilitarian ethic is clearly an extremely demanding ethic in requiring us to promote the general happiness. We can see this in the work of the most famous contemporary utilitarian, Peter Singer. In his well-known article "Famine, Affluence, and Morality," Singer argues for an obligation to help relieve poverty around the world on the basis of the principle that "if it is in our power to prevent something very bad from happening without thereby sacrificing anything morally significant, we ought, morally, to do it."[36] To illustrate the application of this principle, he describes a hypothetical situation where he is walking past a shallow pond and sees a child drowning in it. Singer rightly thinks that he is obligated to wade in to pull the child out, even though this will mean getting his clothes muddy, since muddying one's clothes is morally insignificant, whereas the death of the child would be very bad. While

---

[33] John Stuart Mill, *Utilitarianism* (Indianapolis, IN: Hackett, 2001 [1861]), 17.

[34] Many later utilitarians, such as Peter Singer, would add that we should not show preference for our own species over other sentient species. However, here I will be focusing on our moral responsibilities to other humans.

[35] Bernard Williams, *Moral Luck* (Cambridge: Cambridge University Press, 1981), 18.

[36] Peter Singer, "Famine, Affluence, and Morality," *Philosophy & Public Affairs* 1:3 (1972): 231. Singer in fact endorses a stronger principle, namely: "If it is in our power to prevent something bad from happening, without thereby sacrificing anything of comparable moral importance, we ought, morally, to do it."

all of this seems uncontroversial, Singer says that his principle, in fact, has radical implications:

> If it were acted upon...our lives, our society, and our world would be fundamentally changed. For the principle takes...no account of proximity or distance. It makes no moral difference whether the person I can help is a neighbor's child ten yards from me or a Bengali whose name I shall never know, ten thousand miles away.... If we accept any principle of impartiality...we cannot discriminate against someone merely because he is far away from us.   (231–2)

Singer thinks that in light of this we can see that much of what we spend our money on—from nice clothes for ourselves to birthday presents for our children—is morally unjustified, and we ought instead to donate this money to famine relief organizations around the world. It is worth noting that for strategic reasons Singer, in fact, adopts a weaker principle than the utilitarian principle that we should promote the greatest happiness for the greatest number, which could be even more demanding, even requiring us to sacrifice our own lives if this would promote the general happiness. But Singer's argument is, in any case, in keeping with the spirit of utilitarianism in denying not only the moral significance of our particular attachments to family, friends, and fellow citizens but also the moral significance of proximity.

I want to take issue with both of these points by arguing for neighborliness and loyalty as important limiting virtues. I will start first with the virtue of neighborliness, which I take to be a kind of human solidarity that recognizes the moral significance of proximity. The virtue of neighborliness, as I understand it, is the virtue of being properly responsive to the dignity of other human beings in face-to-face (or close) encounters and the demands they can make on us.[37] The idea of "neighborliness" is perhaps suggestive of someone who is helpful and considerate to those who live nearby, and this is part of what I have in mind, but above all

---

[37] For similar accounts of neighborliness, see Peter Winch, "Who is My Neighbour?," in *Trying to Make Sense* (Oxford: Blackwell, 1987), 154–66, and Jeremy Waldron, "Who Is My Neighbor?: Humanity and Proximity," *The Monist* 86:3 (2003): 333–54.

I have in mind the teachings about the love of neighbor that we find in Judaism and Christianity.

Sometimes these teachings are actually interpreted as offering an impartialist ethic. For instance, immediately after Mill's remarks about the impartiality requirement of utilitarianism, he says that Jesus's teachings about loving your neighbor as yourself "constitute the ideal perfection of utilitarian morality." But the ethic of *neighbor* love is not an impartialist ethic; it recognizes the moral significance of proximity. Indeed, G. K. Chesterton notes that:

> the old religions and the old scriptural language showed so sharp a wisdom when they spoke, not of one's duty towards humanity, but one's duty towards one's neighbour. The duty towards humanity may often take the form of some choice which is personal or even pleasurable.... But we have to love our neighbour because he is *there*—a much more alarming reason for a much more serious operation. He is the sample of humanity which is actually given us. Precisely because he may be anybody he is everybody.[38]

We see this idea of loving or being properly responsive to the one who is *there* when Jesus is asked "Who is my neighbor?" and responds with the Parable of the Good Samaritan, who comes to the aid of a stranger in dire need (Luke 10:30–7). With this parable Jesus clearly means to challenge the conventional idea that our neighbor is just someone who lives nearby and is part of our community. There is a universalization of the ethic of neighbor love in the parable, but it still recognizes the moral significance of proximity, as suggested by the concept of "neighbor" (one who is *nigh*): "when [the Good Samaritan] saw him, he had compassion on him." The message of the parable is that we ought to be ready to act with solidarity with any human being we come across, that is, with "the sample of humanity which is actually given us."[39]

---

[38] G. K. Chesterton, *Heretics* (1905), in *The Collected Works of G. K. Chesterton*, Vol. I (San Francisco, CA: Ignatius Press, 1986), 140.

[39] I have discussed the Parable of the Good Samaritan at greater length in *Virtue and Meaning*, ch. 3. There is some overlap in the discussion here.

The virtue of neighborliness thus calls us to solidarity with concrete humanity, rather than abstract humanity. There is a danger in an abstract love of humanity, which is that in loving humanity in general we may, in fact, love no one in particular, since particular human beings can be difficult to love, given their various imperfections (and our own). We see this expressed by a character in Fyodor Dostoevsky's *The Brothers Karamazov*, who confesses:

> I love mankind... but I am amazed at myself: the more I love mankind in general, the less I love people in particular, that is, individually, as separate persons. In my dreams...I often went so far as to think passionately of serving mankind, and, it may be, would really have gone to the cross for people if it were somehow suddenly necessary, and yet I am incapable of living in the same room with anyone even for two days, this I know from experience.... On the other hand, it has always happened that the more I hate people individually, the more ardent becomes my love for humanity as a whole.[40]

Any authentic love for humanity must be concrete. While we might be able to have a kind of imaginative love for humanity, actual love must be for actual human beings. Without this we can be led to hate actual human beings in pursuit of high-minded ideals, given the ways actual human beings do not live up to our ideals, or we may perhaps come see them as able to be disregarded or even sacrificed for the sake of some supposed general good. It is considerations like these that led William Blake to remark: "He who would do good to another must do it in Minute Particulars: General Good is the plea of the scoundrel, hypocrite & flatterer."[41]

This does not mean that we should not be concerned with people in need around the world, given that we recognize—as the virtue of neighborliness requires—the intrinsic dignity of every human being. Such general concern should arise out of love for concrete humanity. But

---

[40] Fyodor Dostoevsky, *The Brothers Karamazov*, trans. Richard Pevear and Larissa Volokhonsky (New York: Everyman's Library, 1990 [1880]), 57.

[41] William Blake, *Jerusalem: The Emanation of the Giant Albion* (Princeton, NJ: Princeton University Press, 1998 [1805]), 219 (I have added some punctuation).

whereas the obligation of coming to the aid of someone in dire need who is *there* before us is often straightforward (though there could be complicating factors; for instance, if it is a dangerous situation for oneself), how we should act with regard to poverty around the world is much more complicated. As Andrew I. Cohen points out, there is an important difference between an "easy rescue" case (such as Singer presents with his drowning child example) and distant peoples suffering from hunger: "One is an *emergency* calling for immediate action; the other is a *chronic* condition calling for reflection on complex moral, political, and economic considerations."[42] Relatedly, in easy rescue and other local aid cases we can see clearly that we are helping, whereas it is often a matter of dispute whether aid to distant peoples—from governments and individuals donating to NGOs—is overall helpful or harmful to the intended beneficiaries (concerns are raised about how aid can get into the wrong hands and further entrench problematic forms of political and economic power, it can undermine democratic processes and personal initiative, it is often paternalistic and insufficiently attentive to local input, it often lacks a holistic approach, etc.).[43] Even if we can be assured that our aid is helping, world poverty is a large-scale, multifaceted problem that no individual can solve on his or her own, though the super-rich presumably are able to do quite a bit, and everyone who is well-off can do something.

It needs to be noted, however, that the impartiality requirement of utilitarianism in demanding that we always promote the general happiness is, in fact, impossibly demanding—Singer himself does not live up to it—and seeking to follow it is self-mutilating and self-alienating because it requires us to sacrifice the personal projects and relationships that are integral to living a worthwhile human life.[44] Therefore, properly responding to the problem of world poverty will require discretion: we

---

[42] Andrew I. Cohen, "Famine Relief and Human Virtue," in *Contemporary Debates in Applied Ethics*, 2nd edn, ed. Andrew I. Cohen and Christopher Heath Wellman (Malden, MA: Blackwell, 2014), 433.

[43] See Martha C. Nussbaum, *The Cosmopolitan Tradition: A Noble But Flawed Ideal* (Cambridge, MA: Belknap Press of Harvard University Press, 2019), 222–9.

[44] See Bernard Williams, "A Critique of Utilitarianism," in *Utilitarianism: For and Against*, coauthored with J. J. C. Smart (Cambridge: Cambridge University Press, 1973), 95–118; Williams, *Moral Luck*, ch. 1.

have to figure out where we can best help and how to prioritize the different ethical demands that we rightly recognize, both those that stem from recognizing human dignity and those that stem from our particular relationships.

Here we should turn to consider the limiting virtue of loyalty and how it informs the way we think of our duties of assistance. Since the virtue of loyalty is the key virtue that expresses proper partiality, it can, therefore, place limits on how far we can be expected to go in pursuing impartial concern. Loyalty involves a binding attachment to some and not to others, which is maintained through thick and thin. This involves an identification with and a preferential disposition to promote the good of those to whom (or which) one is loyal.[45] The virtue of loyalty expresses such partiality in a proper way and toward proper objects, as a matter of proper responsiveness to intrinsic value (e.g., other human beings and our relationship to them).

In Chapter 1 I discussed the idea of loyalty to the given world (as opposed to other possible worlds), which I suggested provides the wider context for our more particular loyalties. We can, in fact, also speak about a loyalty to humanity (or sentient life), such that what is regarded as impartial morality can be understood as a wider form of partiality. However, here I want to discuss more particular loyalties, especially to friends and family. In Chapter 3 I will consider patriotism, which involves loyalty to one's country and its citizens.

We should first note that there is a connection between the virtue of neighborliness and the virtue of loyalty, namely, when we love and care for those who are *there* in our lives, we will form identity-constituting bonds of attachment with some of these particular people and will come to recognize demands of loyalty to them that sustain the good of the relationship and which give grateful recognition to the good we have received from them. Consider friendship: as Aristotle says, in friendship

---

[45] My view of loyalty here is similar to John Kleinig's; he writes: "[Loyalty] is constituted centrally by perseverance in an association to which a person has become intrinsically committed as a matter of his or her identity" ("Loyalty," *Stanford Encyclopedia of Philosophy*, 2013, http://plato.stanford.edu/entries/loyalty/, accessed July 28, 2021). Later he says: "loyalty can be characterized as a practical disposition to persist in an…associational attachment, where that involves a potentially costly commitment to secure or at least not to jeopardize the interests or well-being of the object of loyalty."

we regard our friends as an extension of ourselves, and we wish and pursue good for our friends as for ourselves and find greater human fulfillment in doing so (see NE IX.4, 9, 12). However, the number of close friends we can have is very limited, since friendship involves a sharing of life together; indeed, Aristotle says: "Those who are many-friended, and treat everyone they meet as if they were their own kin, seem to be friends with no one" (NE IX.10, 1171a15–16). Someone who does not show special regard for a friend over a stranger is no friend at all, and so misses out on a crucially important human good. As Aristotle says: "no one would choose to live without friends, even if he had all the other good things" (NE VIII.1, 1155a4–5). Loneliness—or friendlessness—is one of the worst afflictions from which human beings can suffer, and so a life that eschews binding loyalties to other human beings is hard to recognize as a *human* life, and it certainly cannot be a fulfilled human life, given we are the kind of social creatures that we are.[46] Loyalty thus makes possible crucially important goods of human relationship, and it gives grateful recognition to the goods we have received.

Consider also family relationships (which can be seen as a kind of friendship), starting with marriage, where loyalty involves marital fidelity. At the end of the discussion of temperance in the first section of this chapter I distinguished between lust or mere sexual appetite, which is common in animal life, and romantic love, which is a distinctively human mode of sexual desire. We can add here that romantic love seems to have an inherent "nuptiality": it tends toward and, arguably, is best fulfilled within a permanent and exclusive loving relationship.[47]

---

[46] John Cottingham writes:

> The Aristotelian blueprint for ethical excellence implicitly presupposes, from the outset, a world in which people are already deeply involved in civic and personal networks of partiality.... The opening question for ethics is: how should I—this particular, biologically-based creature—live? And the answer—nobly, harmoniously, with rewarding personal relationships, with graceful and well-ordered habits of desire—makes my own life special and precious in a way which... impersonalist systems of morality cannot properly accommodate.... [The] virtue theorist accepts, and builds upon, the structural constraints of our human nature.
>
> ("Partiality and the Virtues," in *How Should One Live?: Essays on the Virtues*, ed. Roger Crisp [Oxford: Clarendon Press, 1996], 59–61)

[47] See Roger Scruton, *Sexual Desire: A Philosophical Investigation* (New York: Continuum, 2006 [1986]), 339.

The idea here is that when we really love someone romantically, we do not want to live without her or him. Rather, we want to bind our lives together, and this also demands exclusivity as befitting the profound intimacy of the romantic loving relationship. Romantic love, therefore, finds it proper expression and fulfillment in the vow of marriage, and, we should add, this bond can be further enhanced through having children and the making and sustaining of family life. In the vow of marriage, we solemnly promise our love and fidelity to our spouse "for better or for worse," until we are parted by death. Whether or not one is fully convinced by arguments for the claim that romantic love is best fulfilled within monogamous marriage, one can at least see how loyalty in the form of marital fidelity not only helps to foster and protect the distinct human good of the romantic loving relationship but also serves a key social function in providing a stable and loving environment in which to raise children that result from this relationship.[48]

Loyalty to one's own child is also something natural, and it makes possible important human goods. Almost everyone—except apparently Plato (if we take his *Republic* at face value) and a few other radical social reformers[49]—is horrified by the inhumanity of a *Brave New World* type of arrangement where children are raised by the state and there are no special bonds between parent and child. Parents identify with their children in a profound way. As Leon Kass puts it:

Flesh of their flesh, the child is the parents' own commingled being externalized, and given a separate and persisting existence... Providing an opening to the future beyond the grave, carrying not only our seed but also our names, our ways, and our hope that they will surpass us in goodness and happiness, children are a testament to the possibility of transcendence.[50]

---

[48] See Roger Scruton, "Meaningful Marriage," in *A Political Philosophy: Arguments for Conservatism* (London: Bloomsbury, 2006), 81–102.

[49] See Plato, *Republic*, 414c–417b, 423e–424a, 449c–466d; see also Aristotle's critique in *Politics* II.1–4.

[50] Kass, *The Beginning of Wisdom*, 118. Similarly, Aristotle says:

Parents love their children...as they love themselves, since what has come from them is like another (by being separated from them) "themselves"...[Children] seem to be a bond of union, which is why childless unions are more quickly

The loyalty that children owe to their parents is different from the other forms of loyalty discussed so far in this section in that a child's relatedness to his or her parents is not chosen because children do not choose to be conceived. The other relationships, on the other hand, typically do involve some important element of choice (at least in the contemporary world): we choose our friends, we choose our spouse, and we choose to have children (though we do not choose who exactly our children are, unless we pursue genetic engineering). Therefore, the loyalty that children owe to their parents is a matter of filial piety, in line with the discussion in the first section of this chapter. Because our parents give us the gift of life and, when acting as they should, provide us with life-sustaining care and assistance (including character formation) that enables us to mature and achieve our good, we ought to strive to be in right relationship with our parents, which can include, especially as they get older and more dependent, being willing to provide care and support for them, just as they provided for us.[51] Filial piety and loyalty in relation to parents can thus be understood as a duty of gratitude (indeed, I have suggested that loyalty generally expresses gratitude for the goods received from a particular relationship). Such filial piety and loyalty are also important for realizing our human good in that they provide us with a crucially significant form of belonging. Gilbert Meilaender puts the point well:

[We] need to know ourselves as embodied creatures who occupy a fixed place in the generations of humankind. Lines of kinship and descent locate us and identify us, and, unless we learn to accept such a limit on our freedom, we remain alienated from our shared human nature. Such alienation is, at least in part, overcome as we learn to keep

---

dissolved. For children are a good common to both, and what is common holds things together.   (NE VIII.12, 1161b26–8, 1162a26–8)

[51] Aristotle remarks:

[Children] love parents as being what brought them into the world.... The friendship of children towards parents, as of human beings toward gods, is as toward something good and superior, since they are the producers of the greatest goods and the cause of their existence and nurture as well as of their education, once born.

(NE VIII.12, 1161b29, 1162a4–7)

the [biblical] commandment that calls upon us to honor our father and mother...: to show gratitude for a bond in which we find ourselves without ever having chosen it.... We did not create ourselves; we simply find ourselves here.... To learn to affirm and give thanks for our place in lines of kinship and descent is to begin to learn how to give thanks for the mysterious gift of life. We learn to accept and rejoice in our limits, our creatureliness, and we learn gradually to relinquish the secret longing to be more than that. And we learn that before we can love everyone we must accept the hard task of learning to love someone—that is, to love those given to us in the close tie of kinship.[52]

This passage recalls the discussion in Chapter 1 of the importance of an accepting-appreciating stance toward the given, and it also nicely reiterates Chesterton's point that we first need to learn how to love those who are *there* in our lives, starting with family.

In concluding this chapter, we should note some questions that arise about the limits of loyalty. I have been considering how loyalty is a limiting virtue by acknowledging a proper place for partiality in human life and thus placing limits on impartial moral demands. But we can also raise questions about whether and how morality also places limits on loyalty. Concerns are often raised about how loyalty can be used to legitimate bad actions, or at least cover for bad actions.[53] One response—which I think is a good one—is to insist on the unity of the virtues, and so in order for loyalty to be a genuine virtue it has to be compatible with other virtues like justice. However, there are hard cases. Consider, for instance, a case in the *Analects* where the Duke of She tells Confucius that among his people there is someone who is regarded as upright, and when his father stole a sheep, this person reported him to the authorities. Confucius replies: "Among my people, those who we consider 'upright' are different from this: fathers cover up for their sons,

---

[52] Gilbert Meilaender, *Bioethics: A Primer for Christians*, 2nd edn (Grand Rapids, MI: Eerdmans, 2004), 13–14.

[53] See Simon Keller, *The Limits of Loyalty* (Cambridge: Cambridge University Press, 2007); Troy Jollimore, *On Loyalty* (New York: Routledge, 2013), chs. 3–4; and John Kleinig, *On Loyalty and Loyalties: The Contours of a Problematic Virtue* (Oxford: Oxford University Press, 2014), chs. 5 and 7.

and sons cover up for their fathers. 'Uprightness' is found in this"
(13.18).[54] Who is right here? We might find ourselves uncomfortable
with both sides (at the very least with their confidence in where uprightness lies): we do not like the idea of turning in our own father or son to
the authorities, but we also do not like the idea of simply covering for
him. We would like to implore him to do the right thing by making
amends. But perhaps he does not listen. Then what? I will not try to
offer a detailed solution to the matter. However, the fact that most
people would recognize a real dilemma here at least indicates that
most people recognize that loyalty has a legitimate place in human life,
even if it must be weighed against other considerations.

We might also wonder whether loyalty should be regarded as defeasible. What if a friend becomes bad; should we remain friends? A loyal
friend will certainly try to persevere with his or her friend through bad
times and try to help him or her turn things around. But there may come
a point where one can no longer be friends with another person because
they no longer have a shared pursuit of the good, though one will often
still have goodwill toward the other person.[55] Loyalty in family relationships seems less defeasible, and one might argue that it ought to be
regarded as indefeasible. But there are cases of abuse where we rightly
think one should not stay in close relationship, even if familial regard
remains. Again, I will not try to work through all the questions that arise
regarding the defeasibility of loyalty, but most people recognize that
there are real difficulties in these cases, which shows that they recognize

---

[54] See also the similar discussion between Socrates and Euthyphro in Plato's *Euthyphro*.

[55] Aristotle says of those who have the best kind of friendship (friendship in the fullest sense)
based on virtue in the shared pursuit of the good: "their friendship lasts as long as they are
good—and virtue is something steadfast" (NE VIII.3, 1156b11–12). What if our friends become
bad? Aristotle says:

> If they do admit of rectification, we should aid them…with their character…A
> person who did dissolve such a friendship, however, would seem to be doing nothing
> strange, since it was not to a person of that sort that he was a friend. So when his
> friend has changed, and he is unable to restore him to what he was, he disclaims him.
> (IX.3, 1165b18–23)

In cases where a friend does not become worse, but one becomes much more virtuous than his
or her friend, Aristotle thinks "they cannot be friends, since they cannot share a life," but he also
thinks there should be some friendly regard in memory of the former friendship (IX.3,
1165b24–35).

that loyalty has a legitimate place in human life, even if it must be weighed against other considerations.

As previously indicated, in Chapter 3—to which we must now turn—we will have further occasion to examine the place of loyalty in human life, particularly with respect to the question of whether patriotism should be regarded as a genuine virtue.

3

Political Limits

In this chapter I will explore the importance of limits in the political domain, starting with a discussion of the bonds and bounds of political community. I will take on the cosmopolitan outlook that rejects patriotism, which involves loyalty to one's country and fellow citizens. Patriotism, I contend, is important not only because it meets the human need for belonging but also, relatedly, because it enables democratic self-government and distributive justice. In the second section, I will argue against egalitarian views of distributive justice (with a focus on luck egalitarianism) and in favor of a sufficientarian view, where what matters is that everyone has *enough*. This sufficientarian view embraces the limiting virtue of contentment and with this a politics of imperfection, and so in the third section I will argue against utopianism, including what is called "ideal theory" in political philosophy. In the final section I will discuss the importance of the limiting virtue of moderation in politics and, in connection with this, the importance of limiting government. Many of the most important pursuits in life, I contend, lie outside of the political domain, and politics should make space for these pursuits.

The Bonds and Bounds of Political Community

One of the most conspicuous political divides of our age is between those who see themselves primarily as citizens of the world (i.e., cosmopolitans) and those who see themselves primarily as citizens of somewhere in particular. In other words, as David Goodhart puts it, this is a divide between the "anywheres" and the "somewheres."[1] The "anywheres" give

---

[1] David Goodhart, *The Road to Somewhere: The Populist Revolt and the Future of Politics* (London: Hurst, 2017).

The Virtues of Limits. David McPherson. Oxford University Press. © David McPherson 2022.
DOI: 10.1093/oso/9780192848536.003.0004

their primary political allegiance to humanity in general, whereas the "somewheres" endorse patriotism—i.e., loyalty to one's country and its citizens—as a primary political virtue. Without denying the value of our common humanity, I want to take the side of the "somewheres" against the "anywheres" here and defend the importance of patriotism as a genuine virtue. But first we should consider a prominent and influential defense of cosmopolitanism.

In her essay "Patriotism and Cosmopolitanism," Martha Nussbaum draws on ancient Stoic sources to make her case for the cosmopolitan ideal as a way of transcending the divisions and conflicts that exist between people around the world.[2] According to the ancient Stoics:

> [Each] of us dwells, in effect, in two communities—the local community of our birth, and the community of human argument and aspiration that "is truly great and truly common, in which we look neither to this corner nor to that, but measure the boundaries of our nation by the sun"…It is this community that is, fundamentally, the source of our moral obligations…. The accident of birth is just that, an accident; any human being might have been born in any nation. Recognizing this,… we should not allow differences of nationality…to erect barriers between us and our fellow human beings. We should recognize humanity wherever it occurs, and give its fundamental ingredients, reason and moral capacity, our first allegiance and respect.   (7)

Nussbaum notes that the Stoics did not think that we should be devoid of local affiliations, but rather we should see ourselves in "a series of concentric circles," starting with the self and extending out to family,

---

[2] Martha C. Nussbaum, "Patriotism and Cosmopolitanism," in *For Love of Country?*, ed. Joshua Cohen (Boston, MA: Beacon Press, 1996), 3–17. Nussbaum has discussed cosmopolitanism at greater length in her recent book *The Cosmopolitan Tradition* (2019). In this book she says that the cosmopolitan tradition has held a "noble but flawed" ideal in so far as it has rightly recognized the equal dignity of all human beings but has wrongly regarded the dignity of our moral capacity as something complete in itself and so as not requiring material aid (5). She also believes it has failed to recognize the claims of non-human animals, the natural world, and humans with severe cognitive disabilities (2–3, 16–17). These are flaws she seeks to correct, and she also gives more recognition to the significance of particular ties to family, friends, and nations (see chs. 6–7) than in her essay "Patriotism and Cosmopolitanism." I focus on this essay, however, because it provides a concise and influential defense of a strong version of cosmopolitanism.

neighbors, fellow city dwellers, fellow country men and women, and ultimately humanity as a whole, where this outer circle has our primary allegiance (9).[3]

Nussbaum goes on to advocate for making world citizenship, rather than national citizenship, the primary focus of civic education. She makes four points here. First, through cosmopolitan education she believes we can learn more about ourselves: by considering the viewpoints of different people around the world we can "come to see what in our practices is local and nonessential, what is more broadly or deeply shared" (11). Second, through cosmopolitan education we can make progress in solving problems—e.g., environmental problems—that require international cooperation. Third, we "recognize moral obligations to the rest of the world that are real and that otherwise would go unrecognized." Nussbaum does not deny a place for giving one's own particular sphere special concern, but this has to be justified in universalist terms; she writes: "Politics, like child care, will be poorly done if each thinks of herself equally responsible for all, rather than giving the immediate surroundings special attention and care" (13). Finally, Nussbaum notes that through advocating for cosmopolitan education we make "a consistent and coherent argument based on distinctions we are prepared to defend," which are tied to human dignity rather than what is merely contingent or accidental, namely, where one is from, and what she regards as the "morally arbitrary boundary" of the nation (14). Nussbaum concludes on an austere and somber note:

> Becoming a citizen of the world is often a lonely business. It is … a kind of exile—from the comfort of local truths, from the warm, nestling feeling of patriotism, from the absorbing drama of pride in oneself and one's own…. Cosmopolitanism offers no such refuge; it offers only reason and the love of humanity, which may seem at times less colorful than other sources of belonging.   (15)

In Chapter 2 I already called into question the idea of an abstract love of humanity, which poses the danger of loving no one in particular. I made

---

[3] It is worth noting that Stoicism was found attractive in an imperial context.

the case for the limiting virtue of neighborliness as a kind of human solidarity that recognizes the moral significance of proximity. As we saw Chesterton remark, the ethical demand to love (or care for) one's neighbor means that "we have to love our neighbour because he is *there*," that is, "[he] is the sample of humanity which is actually given us." These remarks lend support for seeing patriotism as a virtue: when we love or feel solidarity with those who are there in our lives and with whom we share in political community, it is then appropriate that we form the sort of bonds of loyalty that are part of the virtue of patriotism. As we saw, Nussbaum allows a place for giving one's own particular sphere special concern, but such concern has to be justified in universalist terms, where caring for one's child and one's country is often regarded as the most effective way in which to live out our concern for humanity. But there are also nonderivative justifications for particularistic concern, namely, one's child is *one's own* and one's country is *one's own*, and such particularistic relationships have intrinsic value, and it is morally important to show special concern for those who are *there* in our lives.

Patriotism, it should be noted, is rooted not only in love for (or solidarity with) fellow citizens; it is also rooted in love of place. We can see this in Chesterton's account of the virtue of patriotism, which resembles what he says about love of neighbor with regard to recognizing the significance of proximity:

Patriotism begins the praise of the world at the nearest thing, instead of beginning it at the most distant, and thus it insures what is, perhaps, the most essential of all earthly considerations, that nothing upon earth shall go without its due appreciation. Wherever there is a strangely-shaped mountain upon some lonely island, wherever there is a nameless kind of fruit growing in some obscure forest, patriotism insures that this shall not go into darkness without being remembered in a song.[4]

---

[4] G. K. Chesterton, "The Patriotic Idea" (1904), in *The Collected Works of G. K. Chesterton, Volume XX* (San Francisco, CA: Ignatius Press, 2001), 597.

It is important that human beings have this capacity to love where they are from, to find beauty in the place in which they live, and thereby to belong to and be at home in their place. Human beings have a deep need to belong and to be at home in the world (as I have already touched on in Chapter 1), and patriotism is an important way in which we fulfill this need.[5] True belonging requires the particularity of place. As Mark T. Mitchell notes: "We are limited, placed creatures, and therefore our existence is in some respects necessarily local. If this is the case, then politics, economics, education, and other vital elements of society must be centered on the local and the placed."[6] Mitchell advocates for what he calls "humane localism" as a third alternative in addition to the two alternatives that Nussbaum presents between cosmopolitanism and antagonistic tribalism. Humane localism, he says, "is characterized by a love for one's particular place and the people thereof. Yet at the same time this humane localism is not animated by fear of the other, for by an act of imagination it sees through the inevitable differences and recognizes the common humanity we all share" (100). In Nussbaum's remarks she seems to suggest that whatever is local rather than universal is "nonessential," but most of what we regard as essential to our lives, as that without which we would cease to be ourselves, is, in fact, local and involves particular loves and loyalties.[7] This is something universal about human beings.[8]

---

[5] Simone Weil writes: "To be rooted is perhaps the most important and least recognized need of the human soul.... It is necessary for [one] to draw wellnigh the whole of his moral, intellectual and spiritual life by way of the environment of which he forms a natural part" (*The Need for Roots* [New York: Routledge, 1952], 43). See also Roy Baumeister and Mark Leary, "The Need to Belong: Desire for Interpersonal Attachments as a Fundamental Human Motivation," *Psychological Bulletin* 117:3 (1995): 497–529.

[6] Mark T. Mitchell, "Making Places: The Cosmopolitan Temptation," in *Why Place Matters: Geography, Identity, and Civic Life in Modern America*, ed. Wilfred M. McClay and Ted V. McAllister (New York: Encounter Books, 2014), 96.

[7] Nussbaum does acknowledge that "[we] need not give up our special affections and identifications," and "[we] need not regard them as superficial, and we may think of our identity as partly constituted by them" (9), however, she does not properly recognize how this places considerable constraints on her cosmopolitanism.

[8] Chesterton writes:

> [Real] universality is to be reached rather by convincing ourselves that we are in the best possible relation with our immediate surroundings. The man who loves his own children is much more universal, is much more fully in the general order, than the man who dandles the infant hippopotamus or puts the young crocodile in a perambulator. For in loving his own children he is doing something which is

It is also important to note that cosmopolitanism does not, in fact, overcome "us versus them" thinking. To do so, everyone would have to adopt the cosmopolitan outlook; but clearly not everyone does, and it is highly unlikely that this would ever be the case. As I have already noted, one of the key divides of our age is between cosmopolitan "anywheres" and patriotic "somewheres." Furthermore, those who tend toward the cosmopolitan viewpoint are typically among the highly mobile economic and social-status elites of society, and they often look down on patriotic "somewheres" as being xenophobic and prejudiced, rather than as motivated by a noble and proper love for and loyalty to their place, their culture and traditions, and the people with whom they share these things. Here we are far from transcending "us versus them" thinking, and indeed such thinking is operative in Nussbaum's argument.[9]

I think humane localism is the right approach: we need to live on a human scale as limited, placed creatures who are bound by intrinsically valuable and identity-constituting loves and loyalties, while also recognizing the intrinsic dignity of our common humanity. This approach does not deny the need for forms of governance beyond the local level,

(if I may use the phrase) far more essentially hippopotamic than dandling hippopotami; he is doing as they do. It is the same with patriotism. A man who loves humanity and ignores patriotism is ignoring humanity.   ("The Patriotic Idea," 597)

I think Nussbaum's defense of cosmopolitanism in "Patriotism and Cosmopolitanism" ignores humanity by seeking to transcend it, which elsewhere she has committed herself to avoiding, as in her essay "Transcending Humanity." This represents, I believe, a tension between the Aristotelian and Stoic elements of Nussbaum's thought, and it is not surprising she has abandoned this strong version of cosmopolitanism in her new book (see n. 2). Recall her remarks cited in Chapter 1: "Human limits structure the human excellences, and give excellent action its significance.... [Hubris is] the failure to comprehend what sort of life one has actually got, the failure to live within its limits" ("Transcending Humanity," 378, 381; see also Chapter 2, n. 46). Contrast these remarks with the following from Robert Nozick: "Attempts to find meaning in life seek to transcend the limits of an individual life. The narrower the limits of a life, the less meaningful it is... For a life to have meaning, it must connect with other things, with some things or values beyond itself" (*Philosophical Explanations* [Cambridge, MA: Belknap Press of Harvard University, 1981], 594). I have affirmed that a meaningful life will acknowledge self-transcending sources of value, but what I deny is that a meaningful life is opposed to recognizing limits; indeed, my argument in this book is quite to the contrary, and Nussbaum also agrees in "Transcending Humanity."

[9] I want to thank my student Miranda Hauke, who perceptively made a point along these lines in her final paper on the topic of cosmopolitanism and patriotism for my "Philosophical Ethics" course (Spring 2021), which made me realize that this is a point that I should also make here. On these matters, see Christopher Lasch, *The Revolt of the Elites and the Betrayal of Democracy* (New York: Norton, 1996); Michael Lind, *The New Class War: Saving Democracy from the Managerial Elite* (New York: Portfolio, 2020); Michael J. Sandel, *The Tyranny of Merit: What's Become of the Common Good?* (New York: Farrar, Straus and Giroux, 2020).

including at the national and international level. But it embraces the *principle of subsidiarity*, which says that we ought to deal with political matters as close to the matter at hand as possible, and higher levels of government should support the lower levels. The concern here is with facilitating democratic self-government. We can see the principle of subsidiarity operative in the system of federalism established by the US Constitution, which delineates the specific powers of the federal government and leaves the rest to state and local governments and to individuals and their communities (as specified by the Tenth Amendment).[10] This involves a commitment to political decentralization or dispersed sovereignty. However, over time there, in fact, has been an increasing centralization of political life, which—along with the globalization of the economy—has posed significant challenges for democratic self-government.

One of the key arguments against cosmopolitanism and in favor of patriotism is based on the concern to foster democratic self-government. It is worth considering here Michael Sandel's argument in his book *Democracy's Discontent*. In Chapter 1 we saw that in *The Case against Perfection* Sandel is concerned to combat the Promethean ideal of mastery in genetic engineering, which he thinks goes too far by undermining a sense of life as a gift. However, in *Democracy's Discontent* he is, in fact, concerned with a loss of mastery, that is, of democratic self-government. He notes that there is a widespread fear that, individually and collectively, "we are losing control of the forces that govern our lives," which he sees as also connected with a concern about the erosion of forms of community that "situate people in the world and provide a source of identity and belonging."[11] According to the classical republican political theory (in the tradition of Jefferson and Tocqueville) that Sandel endorses, "to be free is to share in governing a political community that controls its own fate. Self-government in this sense requires political communities that control their destinies, and citizens who identify sufficiently with those communities to think and act with a view to the

---

[10] See Mark T. Michell, *The Politics of Gratitude: Scale, Place & Community in a Global Age* (Washington DC: Potomac Books, 2012), 93–8.

[11] Michael J. Sandel, *Democracy's Discontent: America in Search of a Public Philosophy* (Cambridge, MA: Belknap Press of Harvard University, 1996), 3, 294.

common good," which requires the cultivation of civic virtue (274; see 5–6). Sandel thinks this was realized to a great extent in the early days of the American republic, where political life was decentralized and there was a focus on local democratic engagement. But he traces how, as economic power grew with industrial capitalism, the power of the federal government grew with it,[12] and this also gave rise to what he calls "the procedural republic," which seeks to be neutral between competing conceptions of the good for human life and offers a framework of rights for individuals to pursue their particular conception of the good life (John Rawls has provided the most influential defense of this political order[13]). The procedural republic offers a new understanding of freedom, where "our liberty depends not on our capacity as citizens to shape the forces that govern our collective destiny but rather on our capacity as persons to choose our values and ends for ourselves" (275). However, Sandel contends that "[the] triumph of the voluntarist conception of freedom coincided with a growing sense of disempowerment" (294), as it is unable to master the forces that govern our lives, especially the economic forces that operate on a global scale.

To address democracy's discontent, Sandel believes that we need "political institutions capable of governing the global economy," while at the same time we need "to cultivate the civic identities necessary to sustain those institutions, to supply them with the moral authority they require" (338). However, Sandel does not think cosmopolitanism is up to the task, saying:

> The love of humanity is a noble sentiment, but most of the time we live our lives by smaller solidarities. This may reflect certain limits to the bounds of moral sympathy. More important, it reflects the fact that we learn to love humanity not in general but through its particular expressions.... To affirm as morally relevant the particular communities that locate us in the world, from neighborhoods to nations, is not

---

[12] There were other contributing factors to the growth of the power of the federal government in the United States, such as post-Civil War reconstruction, the Great Depression, the two world wars, and the Civil Rights movement. See Mitchell, *The Politics of Gratitude*, 93.

[13] See John Rawls, *Political Liberalism*, expanded edn (New York: Columbia University Press, 2005 [1993]). I will return to discuss Rawls's work further in the next section.

to claim that we owe nothing to persons as persons, as fellow human beings. At their best, local solidarities gesture beyond themselves toward broader horizons of moral concern, including the horizon of our common humanity. The cosmopolitan ethic is wrong, not for asserting that we have certain obligations to humanity as a whole but rather for insisting that the more universal communities we inhabit must always take precedence over more particular ones.   (342–3)

According to Sandel, the large-scale nation-state now also has difficulty commanding the allegiance of its citizens: "Beset by the integrating tendencies of the global economy and the fragmenting tendencies of group identities, nation-states are increasingly unable to link identity and self-rule" (344). Thus, he thinks that the best hope for self-government is through dispersing sovereignty: "Only a regime that disperses sovereignty both upward and downward can combine the power required to rival global market forces with the differentiation required of a public life that hopes to inspire the reflective allegiance of its citizens" (345). In our age, Sandel says, "the politics of neighborhood matters more, not less" (346):

People will not pledge allegiance to vast and distant entities, whatever their importance, unless those institutions are somehow connected to political arrangements that reflect the identity of the participants.... [Proliferating] sites of civic activity and political power can serve self-government by cultivating virtue, equipping citizens for self-rule, and generating loyalties to larger political wholes.   (346, 348)

One way to think about patriotism in the form of loyalty to a large-scale nation-state, then, is in terms of how it makes possible our particular self-governed ways of life and belonging. For instance, a Midwestern American from Nebraska can feel patriotic loyalty toward the United States for providing a constitutional order that makes possible her or his particular regional way of life. Here patriotism is analogous to filial piety (indeed, the term "patriotism" suggests the idea of one's country as a kind of parent), since one's country is in a sense a source of one's life—at least one's way of life—in providing basic rights and liberties as well as protections from harm (including police and military protection and social safety nets, such as welfare assistance aimed at meeting vital need), and this implies certain responsibilities to give back to one's country, which can be understood as duties of gratitude.[14] A nation will, of course, typically also have a common language and to some degree a common culture informed by a story about the nation's history, which will inform a sense of national loyalty and belonging. However, in many cases these sources of national belonging are losing some of their unifying force, as nation-states are becoming more internally fragmented, as Sandel notes, and also as some people are identifying with a cosmopolitan stance. This poses a challenge to democratic self-government at the nation-state level, and hence the importance of decentralizing or "dispersing sovereignty." But I am suggesting—in line with Sandel—that the nation-state still has a role to play here in facilitating more local forms of democratic self-government and so can command our loyalty.

It is important to note here that I am defending a conception of patriotism as a virtue, but there are also non-virtuous forms of patriotism, or we might say false forms, if we use patriotism to designate the virtuous form of loyalty to one's country and fellow citizens. As discussed in Chapter 2, there are real concerns about the way in which loyalty can be used to legitimate bad actions or at least cover for them, and this must be regarded as incompatible with virtuous loyalty. On this point, Chesterton writes:

"My country, right or wrong," is a thing that no patriot would think of saying except in a desperate case. It is like saying, "My mother, drunk or sober." No doubt if a decent man's mother took to drink he would share her troubles to the last; but to talk as if he would be in a state of gay indifference as to whether his mother took to drink or not is certainly not the language of men who know the great mystery.[15]

---

[14] For Aquinas, patriotism is, in fact, part of the virtue of piety along with filial piety, whereas reverence for God is understood as the virtue of religion, but all concern what is owed to the "principles of our being and government" (*Summa Theologiae* II–II, q. 101, a. 1, trans. Fathers of the English Dominican Province). On my view, patriotism involves a loyalty to one's country and fellow citizens that is properly grounded in both love and justice, where piety or reverence for the sources of one's being is understood as an aspect of justice.

[15] G. K. Chesterton, *The Defendant* (Mineola, NY: Dover [2012 (1902)]), 79.

Patriotism is fully compatible with criticism of one's country and working for improvement; if we love and are loyal to something (or someone), then we want what is good for it (or him or her), and we will work to bring about this good. Moreover, though I agree with Chesterton that decency requires we share our mother's troubles, and likewise we should share in our country's troubles to a great extent, nevertheless, loyalty to one's country can be defeasible if the country becomes pervasively evil and oppressive and one is not able to work for reform.

Let us return to Nussbaum's ideal of world citizenship. There is, of course, no world-state of which one can be a citizen. Therefore, "world citizenship" must be understood as a metaphor that expresses a cosmopolitan view of the primary object of moral and political allegiance. Cosmopolitans do not generally argue for a world-state, which would be hard to reconcile with a commitment to democratic self-government. They usually accept the nation-state in some form, while advocating for strong forms of international governance. The differences between the cosmopolitan "anywheres" and the patriotic "somewheres" often come out, instead, in disputes over immigration, trade, foreign policy, and the kind and extent of international governance.

Consider the issue of immigration. Cosmopolitan "anywheres" are typically in favor of much looser restrictions on immigration than patriotic "somewheres," and, in fact, there seems to be increasing support for open borders, or at least what is basically their equivalent. One common argument for open borders (or something close to them), which is provided by Chandran Kukathas, is what we might call the humanitarian argument. As Kukathas puts it:

> The great majority of the people of the world live in poverty, and for a significant number of them the most promising way of improving their condition is to move.... To say to such people that they are forbidden to cross a border in order to improve their condition is to say to them that it is justified that they be denied the opportunity to get out of poverty...A principle of humanity suggests that very good reasons must be offered to justify turning the disadvantaged away.[16]

---

[16] Chandran Kukathas, "The Case for Open Immigration," in *Contemporary Debates in Applied Ethics*, 2nd edn, ed. Andrew I. Cohen and Christopher Heath Wellman (Malden, MA: Blackwell, 2014), 380.

Kukathas also adds to this an argument about the overall benefits of global free trade and the free movement of people, while acknowledging that some people will lose out in this arrangement, especially, we can add, those already less well off in the receiving country.

So, one of the issues here is whether a country can be justified in protecting its own citizens from negative economic consequences that would attend open immigration. This issue is also closely connected with whether a standard welfare system—which provides healthcare, education, unemployment relief, and disability benefits, and ensures a minimum wage for those who do work—is justified. Kukathas notes that open immigration is a problem for a welfare state; indeed, they do not appear to be compatible, given that a country would not be able to provide adequate welfare assistance to everyone in a system of open immigration. For many people this is a reason to limit immigration; however, for Kukathas, who is a libertarian, it is a reason to reject the welfare state, and indeed the state as we know it, given his commitment to what he calls a "principle of freedom" and a "principle of humanity." To side with limits on immigration, he thinks, is to say that a "principle of nationality" should take precedence, that is, we should give precedence to the needs of our fellow citizens (382–3; see 388).

I think it is right that our fellow citizens should generally take precedence, though—as I will make clear—this does not deny that we should have concern for common humanity. As Michael Walzer writes:

The idea of distributive justice presupposes a bounded world within which distributions take place: a group of people committed to dividing, exchanging, and sharing social goods, first of all among themselves.... The primary good that we distribute to one another is membership in some human community. And what we do with regard to membership structures all our other distributive choices: it determines with whom we make those choices, from whom we require obedience and collect taxes, to whom we allocate goods and services.[17]

---

[17] Michael Walzer, *Spheres of Justice: A Defense of Pluralism and Equality* (New York: Basic Books, 1983), 31.

In other words, distributive justice depends upon patriotism or loyalty to fellow citizens. Walzer goes on to note that what is at stake in a country's immigration policy is:

> the shape of the community that acts in the world…Admission and exclusion are at the core of communal independence. They suggest the deepest meaning of self-determination. Without them, there could not be *communities of character*, historically stable, ongoing associations of men and women with some special commitment to one another and some special sense of their common life…. [It] is only as members somewhere that men and women can hope to share in all the other social goods—security, wealth, honor, office, and power—that communal life makes possible. (61–3).

Building on Walzer's remarks, David Miller argues—in line with the argument from Sandel that we have considered—that states "require a common public culture that in part constitutes the political identity of their members, and that serves valuable functions in supporting democracy and other social goals."[18] In particular, it helps to enable the sort of solidarity (i.e., patriotic loyalty) that makes distributive justice within a welfare system possible. Miller acknowledges that cultures are always changing and there are limits to what we can do to control cultural change; nevertheless, he thinks that part of democratic self-government is cultural self-determination:

> [The] public culture of their country is something that people have an interest in controlling: they want to be able to shape the way that their nation develops, including the values that are contained in the public culture…. [They have reason] to try to maintain cultural continuity over time, so that they can see themselves as the bearers of an identifiable cultural tradition that stretches backward historically. (370)

---

[18] David Miller, "Immigration: The Case for Limits," in *Contemporary Debates in Applied Ethics*, 2nd edn, ed. Andrew I. Cohen and Christopher Heath Wellman (Malden, MA: Blackwell, 2014), 369.

Kukathas thinks that this argument about the importance of maintaining solidarity through a shared public culture in order to support distributive justice within a welfare system is the most challenging argument against open immigration. However, we have already seen that he thinks it is best to reject such a welfare system in favor of open immigration. Kukathas does not think that the nation-state is "the appropriate site for the settlement of questions of distributive justice."[19] For one thing, he thinks that some nation-states may be too large to have a shared conception of distributive justice. But even if nation-states were plausible sites of distributive justice, Kukathas thinks there is another problem:

> Is it right that the preservation of local institutions of social justice take precedence over the humanitarian concerns that make open immigration desirable?...If the price of social justice is the exclusion of the worst-off from the lands that offer the greatest opportunity, this may be a mark against the ideal of social justice.[20]

Kukathas's remarks here seem to make impermissible any form of partiality—whether for family, friends, or fellow citizens—and so these remarks are subject to the same sorts of criticisms that I discussed in Chapter 2 regarding strict impartiality, especially how it alienates us from our own humanity and the sources of human fulfillment. It should also be noted that defending limits on immigration is not incompatible with humanitarian concern. Indeed, both Walzer and Miller acknowledge that a nation-state needs to acknowledge the claims of dire need with regard to refugees in a just immigration policy. However, the claims of dire need around the world need to be weighed against the just claims of fellow citizens. When Kukathas questions whether the nation-state is the appropriate site for settling questions of distributive justice, it seems that "distributive justice" for him just means leaving things to be sorted out by the "invisible hand" of global capitalism. But within this system (as within any human system) people will encounter misfortune, and it seems any humane way of dealing with such

[19] Kukathas, "The Case for Open Immigration," 384.
[20] Kukathas, "The Case for Open Immigration," 385.

misfortune will require some sort of welfare system or social safety net, and it is not clear how we can provide this without recognizing the bonds and bounds of political community. Indeed, as suggested by Aristotle, political community, as a community (*koinōnia*) of sharing within a common location involving bonds of friendship and justice, "comes to be for the sake of living, but it remains in existence for the sake of living well" (P I.2, 1252b26–9; see also P III.9, 1280b29–1181a7; NE VIII.1, 1155a24–30; VIII.9, 1159b25–32; VIII.11).

Kukathas is correct to note the challenges for arriving at a shared understanding of distributive justice within large nation-states, especially ones characterized by multiculturalism. This is one reason why I suggested we need decentralization and to affirm the principle of subsidiarity. However, dealing with disagreement and trying to find some majoritarian consensus are part of democratic political life, and certainly Kukathas's own proposal for open borders and the abolition of the welfare system would meet with widespread disagreement. As best we can, we must draw on and maintain whatever shared sense of identity and belonging that we have in order to support the kind of solidarity that makes democratic self-government and distributive justice possible within the nation-state. I will turn now to make a case for a particular conception of distributive justice within the nation-state.

Sufficientarian Justice

I want to make the case for a *sufficientarian* conception of distributive justice, which I will contrast with and defend against a common *egalitarian* conception known as "luck egalitarianism." The basic sufficientarian claim is that what matters with regard to distributive justice is, as Harry Frankfurt puts it, "not that everyone should have *the same* but that each should have *enough*,"[21] where what counts as "enough," on my view, is that each person can live and live well in a characteristically human mode of life, and where addressing dire human need is of

---

[21] Harry Frankfurt, "Equality as a Moral Ideal," *Ethics* 98:1 (1987), 21.

paramount importance.[22] Indeed, many times when people say they are concerned with inequality, it seems that what they are really concerned with is poverty, that is, that people do not have enough. After all, it does not seem important that two wealthy people have the same amount of resources, but it is important that everyone has enough to live well.

It should be noted up front that although I reject luck egalitarianism as the right form of distributive justice, I do affirm moral egalitarianism, which maintains that all human beings in virtue of being human possess the same basic intrinsic dignity, and which is also connected to a relational egalitarianism where we treat other human beings as our equals with respect to basic dignity. Such moral and relational egalitarianism, I maintain, does not require that everyone should have the same material conditions, but it does require that everyone should have enough, since to disregard dire human need is to disregard human dignity. Where one cannot meet this need for oneself, the obligation falls to others, especially the political community, which I suggested should be understood as in part constituted for this purpose. Beyond meeting dire need, we should also seek to promote a broad "equality of condition" (rather than equality of outcome), which Tocqueville thought was the characteristic condition of genuinely democratic societies.[23] As Sandel describes it, such equality of condition "enables those who do not achieve great wealth or prestigious positions to live lives of decency and dignity—developing and exercising their abilities in work that wins social esteem, sharing in a widely diffused culture of learning, and deliberating with their fellow citizens about public affairs."[24]

With these preliminary remarks now complete, let us now examine luck egalitarianism as the major egalitarian rival to sufficientarianism. Luck egalitarianism finds a prominent supporter in G. A. Cohen, whose essay "Rescuing Conservatism: A Defense of Existing Value" I drew on in

---

[22] I am thinking of distributive justice here primarily in terms of how we distribute economic resources. But if we think of distributive justice as also concerning basic rights and liberties (as key aspects of political membership), then with regard to such rights and liberties having enough would be having the same.

[23] See Alexis de Tocqueville, *Democracy in America*, trans. Harvey C. Mansfield and Delba Winthrop (Chicago: University of Chicago Press, 2002 [1835/1840]), where equality of condition is discussed throughout.

[24] Sandel, *The Tyranny of Merit*, 224.

Chapter 1 in articulating the importance of an accepting-appreciating stance toward the given world. I will argue that luck egalitarianism is in fact at odds with a proper accepting-appreciating stance toward the given, since, as David Wiggins puts it, it wages "a metaphysical crusade against contingency."[25] By contrast, sufficientarianism does not have this problem, and this is an important reason in its favor, as I will return to discuss.

In his essay "On the Currency of Egalitarian Justice," Cohen expresses the basic idea of luck egalitarianism when he writes: "a large part of the fundamental egalitarian aim is to extinguish the influence of brute luck on distribution...Brute luck is an enemy of just equality, and, since effects of genuine choice contrast with brute luck, genuine choice excuses otherwise unacceptable inequalities."[26] In other words, inequalities in society that result from brute luck (i.e., luck due to the contingencies of our natural and social circumstances) are unjust and should be redressed, whereas inequalities that result from genuine choice are not unjust and do not call for redress.

Consider also John Rawls, who is the most influential political philosopher of the last century.[27] According to Rawls, "the accidents of natural endowment and the contingencies of social circumstance" are "arbitrary from a moral point of view" and should not be allowed to determine social and economic advantage.[28] Thus, in order to come up with the fair terms of social cooperation—what Rawls regards as the principles of

---

[25] David Wiggins, *Ethics: Twelve Lectures on the Philosophy of Morality* (Cambridge, MA: Harvard University Press, 2006), 306.

[26] G. A. Cohen, "On the Currency of Egalitarian Justice," *Ethics* 99:4 (1989): 931.

[27] Elizabeth Anderson, who coined the term "luck egalitarianism," regards Rawls as sharing her "relational egalitarianism," though she acknowledges luck egalitarianism can be traced to his work (see "What Is the Point of Equality?," *Ethics* 109:2 [1999]: 290; see also Anderson, "The Fundamental Disagreement between Luck Egalitarians and Relational Egalitarians," *Canadian Journal of Philosophy*, Supplemental Volume 36 [2010]: 1–23). Anderson writes:

> The proper negative aim of egalitarian justice is not to eliminate the impact of brute luck from human affairs, but to end oppression, which by definition is socially imposed. Its proper positive aim is...to create a community in which people stand in relations of equality to others.   ("What Is the Point of Equality?," 288–9)

I think Rawls should be regarded as a luck egalitarian because he is concerned to correct the influence of brute luck on distribution, even if it is right also to regard him as a relational egalitarian.

[28] John Rawls, *A Theory of Justice*, revised edn (Cambridge, MA: Harvard University Press, 1999 [1971]), 14; see also 62–5, 86–90, 273–4.

justice—we need to think of ourselves in the hypothetical scenario of the "original position," where we adopt a "veil of ignorance" such that we do not know the particularities of our social situation, our natural assets, and our conception of the good for human life (10–11). This is supposed to ensure that our choice of the principles of justice is not biased by the contingencies of our natural and social circumstances. There are two general principles of justice that Rawls thinks we will arrive at here. First: "Each person is to have an equal right to the most extensive total system of equal basic liberties compatible with a similar system of liberty for all." Second: "Social and economic inequalities are to be arranged so that they are both: (a) to the greatest benefit of the least advantaged…and (b) attached to offices and positions open to all under conditions of fair equality of opportunity" (266). Part (a) of the second principle expresses "the difference principle," which requires equal distribution of economic benefits unless it can be shown that an unequal distribution is more beneficial to the least well off. Rawls writes:

> The difference principle represents, in effect, an agreement to regard the distribution of natural talents as in some respects a common asset and to share in the greater social and economic benefits made possible by the complementarities of this distribution. Those who have been favored by nature, whoever they are, may gain from their good fortune only on terms that improve the situation of those who have lost out…. No one deserves his greater natural capacity nor merits a more favorable starting place in society. But, of course, this is no reason to ignore, much less to eliminate these distinctions. Instead, the basic structure can be arranged so that these contingencies work for the good of the least fortunate. Thus we are led to the difference principle if we wish to set up the social system so that no one gains or loses from his arbitrary place in the distribution of natural assets or his initial position in society without giving or receiving compensating advantages in return. (87)

Although they are both concerned to correct the influence of brute luck on distribution, Cohen takes issue with the way in which Rawls seeks to incentivize those who are most talented to work for the good of others when he says that they "may gain from their good fortune only on terms that improve the situation of those who have lost out." Cohen thinks we should be motivated by an egalitarian ethos even in our personal decisions, such as in how we live our economic lives. Rawls locates the action of distributive justice in the state, and then leaves individuals to pursue whatever conception of the good life they happen to have within the framework of rights and liberties that the state makes possible. Cohen questions this, as he thinks we need to act as egalitarians in all aspects of our lives; otherwise, we will end up undermining the kind of solidarity that an egalitarian political community requires.[29] As he puts it in the title of one of his books: "If you're an egalitarian, how come you're so rich?"[30]

In another way Rawls is in fact much more radical than Cohen because he rejects the traditional common-sense conception of justice as desert. Not only does he think that we do not deserve our natural capacities or our favorable starting place in society (rightly so, since how could these be deserved?), but he goes further than Cohen in denying desert for individual initiative and responsible choice (recall that Cohen says: "since effects of genuine choice contrast with brute luck, genuine choice excuses otherwise unacceptable inequalities"). Rawls writes: "Even the willingness to make an effort, to try, and so to be deserving in the ordinary sense is itself dependent upon happy family and social circumstances," and so "[the] notion of desert does not apply here."[31] Robert Nozick aptly remarks:

> This line of argument can succeed in blocking the introduction of a person's autonomous choices and actions (and their results) only by attributing *everything* noteworthy about a person completely to certain sorts of "external" factors. So denigrating a person's autonomy and prime responsibility for his actions is a risky line to take for a theory

---

[29] See G. A. Cohen, *Rescuing Justice and Equality* (Cambridge, MA: Harvard University Press, 2008).

[30] G. A. Cohen, *If You're an Egalitarian, How Come You're So Rich?* (Cambridge, MA: Harvard University Press, 2000).

[31] Rawls, *A Theory of Justice*, 64, 89; see also 273–4.

that otherwise wishes to buttress the dignity and self-respect of autonomous beings.[32]

Nozick also takes issue with Rawls's view that we should regard our natural assets as "common assets." He turns Rawls's own argument against utilitarianism against him and contends that he "does not take seriously the distinction between persons" (228). Indeed, on the basis of the atomistic or unencumbered conception of the self that both Rawls and Nozick affirm it is not clear how the difference principle can avoid the charge of violating the Kantian principle that persons should not be used as a mere means to others' ends. As Sandel remarks: "What the difference principle requires, but cannot provide, is some way of identifying those among whom the assets I bear are properly regarded as common, some way of seeing ourselves as mutually indebted and morally engaged to begin with."[33] Short of either a socially encumbered conception of the self in which we can regard the good of others as our own or a conception of desert by which we can be said to be morally obligated to help those who are less fortunate, it is not clear that our assets should be regarded as, in some sense, common assets.

What I want to focus on, however, is the attitude toward contingency, or "the given," in luck egalitarianism. We have seen that Rawls regards the contingencies of our natural and social circumstances as "arbitrary from a moral point of view." But is this right? I do not think so. For one thing, these contingencies of natural and social circumstance, which give rise to particular loyalties, shape our very character as "encumbered selves" and give our lives a "moral depth."[34] As George Sher says, "the innumerable contingencies that differentiate each person's situation from those of others" are not simply "so many sources of unjust inequality to be neutralized by society," but rather they are "the backdrop in whose absence we could not live recognizably human lives at all."[35]

---

[32] Robert Nozick, *Anarchy, State, and Utopia* (New York: Basic Books, 1974), 214.

[33] Michael J. Sandel, "The Procedural Republic and the Unencumbered Self," *Political Theory* 12:1 (1984), 90.

[34] See Sandel, "The Procedural Republic and the Unencumbered Self," 90–1.

[35] George Sher, *Equality for Inegalitarians* (Cambridge: Cambridge University Press, 2014), viii.

David Wiggins also writes in response to Rawls's attempt to "remove the 'undeserved' or 'morally arbitrary' contingencies of natural gifts and fortunate circumstances":

> Even behind the veil of ignorance, the deliberators might easily see the disadvantage of insisting on the 'moral arbitrariness' of such blessings—if insisting on it will have the effect of subverting the message of Jesus' parable of the talents, namely one's paramount moral duty to make something (and not just for oneself) of what gifts one has (Matthew 25: 14–30). Contingency is not the same as arbitrariness.[36]

In other words, the contingencies of our natural and social circumstances are *morally charged*: we ought to do the best with what we have been given, and, as Jesus goes on to instruct, we ought to show a special solicitousness toward those who are in need (Matthew 25:34–46). And such solidarity is also encouraged by being mindful of how others' misfortune could have been (or could be) our own.[37]

In an arresting remark, which I appealed to at the beginning of this section, Wiggins describes Rawls as waging "a metaphysical crusade against contingency," that is, against the given, and I think the same can be said of other luck egalitarians, including Cohen.[38] And thus Cohen's luck egalitarianism stands at odds with his small-c conservatism, or what I called "existential conservatism," which is supposed to encourage a proper accepting-appreciating stance toward the given world (while allowing for improvement). In "Rescuing Conservatism," Cohen does acknowledge some tension between his conception of justice and his existential conservatism, though he maintains: "you can be both egalitarian and conservative by putting justice lexically prior to (other)

---

[36] Wiggins, *Ethics*, 306.

[37] On this point, see Alasdair MacIntyre, *Dependent Rational Animals: Why Human Beings Need the Virtues* (Chicago: Open Court, 1999), 100–1; Sandel, *The Tyranny of Merit*, 25.

[38] Elizabeth Anderson similarly remarks that luck egalitarians are concerned with "correcting a supposed cosmic injustice" ("What Is the Point of Equality?," 288). This is well illustrated when Thomas Nagel, another luck egalitarian, asks: "How could it not be an evil that some people's life prospects at birth are radically inferior to others?" (*Equality and Partiality* [Oxford: Oxford University Press, 1991], 28).

value," that is, by giving justice precedence over other values when they come into conflict. However, he also says: "I do not say that I am myself so uncompromising an egalitarian, so lexically projustice. I am not sure that we should regret the production of all the wonderful material culture that we have inherited and that was produced at the expense of gross injustice" (172). In "On the Currency of Egalitarian Justice," he writes: "if priority were always given to relieving misery, then no resources could be devoted to maintaining cathedrals and other creations of inestimable value." Since human beings have not only material needs but also what we might call spiritual needs—i.e., needs for beauty and other things of higher value that can make us glad to be alive and enable us to affirm life in the world as worthwhile—any position on distributive justice that disregards these spiritual needs is for that reason objectionable. Cohen thinks this is a powerful objection, though he says it "does not challenge the claim that, to the extent that equalization is defensible, welfare is the right thing to equalize" (910–11).[39] But "relieving misery" is not the same thing as equalizing welfare, and if our concern is with the former rather than the latter (as I think it should be), then there is no inherent tension with promoting high culture (e.g., in building and maintaining cathedrals, art museums, universities, etc.). Moreover, the aim of relieving misery is fully compatible with existential conservatism, whereas luck egalitarianism is not because of its "metaphysical crusade against contingency," and this is a point that Cohen does not seem to appreciate.

Luck egalitarianism encourages a continual disappointment with and repudiation of the given world. It gives emphasis to indignation rather than appreciation of the given, and its focus on what it regards as the wrong of economic inequality encourages constant comparison with others, which can lead to feelings of discontent, resentment, and envy that are opposed to the virtues of gratitude and contentment.[40] Such

---

[39] Cohen thinks that the "currency" of egalitarian justice is "access to advantage," where this is "understood to include, but to be wider than, welfare" ("On the Currency of Egalitarian Justice," 907).

[40] See Roberts, "The Blessings of Gratitude" for an illuminating discussion of how gratitude counteracts envy, resentment, and regret. Regarding luck egalitarianism and discontent, Frankfurt writes:

> Nagel maintains that what underlies the appeal of equality is an "ideal of acceptability to each individual." On his account, this ideal entails that a reasonable person

comparison also distracts us from concern for what really matters for one's own life.[41] And yet we find Ronald Dworkin—who is another prominent luck egalitarian—putting forward an "envy test" as a criterion of justice in distribution, writing: "No division of resources is an equal division if, once the division is complete, [a person] would prefer some-one's bundle of resources to his own bundle."[42] We know something has gone drastically wrong when a vice has been elevated to a standard of justice.[43] Luck egalitarians also typically support genetic engineering of children—as we have seen with Dworkin in Chapter 1—and see it as a matter of justice, which perhaps most clearly expresses their "metaphysical crusade against contingency."[44] Cohen opposes such genetic

should consider deviations from equality to be acceptable only if they are in his interest in the sense that he would be worse off without them. But a reasonable person might well regard an unequal distribution as entirely acceptable even though he did not presume that any other distribution would benefit him less. For he might believe that the unequal distribution provided him with quite enough, and he might reasonably be unequivocally content with that, with no concern for the possibility that some other arrangement would provide him with more. It is gratuitous to assume that every reasonable person must be seeking to maximize the benefits he can obtain, in a sense requiring that he be endlessly interested in or open to improving his life.... [Contentedness] precludes having an *active interest* in getting more.... Some people feel that their lives are good enough, and it is not important to them whether their lives are as good as possible.

("Equality as a Moral Ideal," 36, 39)

[41] Frankfurt writes:

The mistaken belief that economic equality is important in itself leads people to detach the problem of formulating their economic ambitions from the problem of understanding what is most fundamentally significant to them. It influences them to take too seriously, as though it were a matter of great moral concern, a question that is inherently rather insignificant and not directly to the point, namely, how their economic status compares with the economic status of others. In this way the doctrine of equality contributes to the moral disorientation and shallowness of our time.   ("Equality as a Moral Ideal," 23)

[42] Ronald Dworkin, *Sovereign Virtue: The Theory and Practice of Equality* (Cambridge, MA: Harvard University Press, 2000), 67.

[43] John Kekes writes in response to Dworkin:

It should not escape notice how extraordinary it is to make envy the test of ideal distribution. Envy is the vice of resenting the advantages of another person. It is a vice because it tends to lead to action that deprives people of advantages they have earned by legal and moral means. The envy test does not ask whether people are entitled to their advantages; it asks whether those who lack them would like to have them.... Instead of recognizing that envy is wrong, Dworkin elevates it into a moral standard.
(*The Illusions of Egalitarianism* [Ithaca, NY: Cornell University Press, 2003], 71)

[44] See Rawls, *A Theory of Justice*, 92.

engineering on the basis of his existential conservatism, but again here we can see a conflict with his luck egalitarianism.

I think in the conflict between existential conservatism and luck egalitarianism we should side with the former over the latter. This is because of what I have argued in Chapter 1 is the spiritual importance of existential conservatism for addressing the problem of cosmodicy, and because luck egalitarianism is not a compelling account of distributive justice. We do better to preserve a traditional common-sense notion of justice as desert, where people are thought to be deserving of certain consideration and treatment in virtue of their basic human dignity, in virtue of the choices they make and who they become, and in virtue of their membership in a political community. And what deserves our special consideration is dire human need. In regard to responding to contingencies of natural and social circumstance, Wiggins writes that the key issue can be stated as follows:

> [Given] that, whatever principles may be instituted by human beings to regulate the social and political spheres, the human world will always be replete with contingency, good luck, bad luck, and the rest, what guarantees of what strength must we place among the conditions of our cooperation in order to ensure that the worst of the bad luck that anyone encounters will be alleviated, along with its consequences, by concerted social action? After all, the realist will say, the first and foremost thing that affects and harms the dispossessed or destitute is dire, unsatisfied need.[45]

In other words, we should be sufficientarians rather than egalitarians. As stated earlier, this means that what matters from the standpoint of distributive justice is that everyone has *enough*, where what counts as "enough," on my view, is that each person can live and live well in a characteristically human mode of life, and where relieving misery—i.e., meeting dire human need—is of paramount importance.[46]

---

[45] Wiggins, *Ethics*, 200. See also David Wiggins, "Claims of Need," in *Needs, Values, Truth*, 3rd edn (Oxford: Clarendon Press, 1998 [1987]), 1–57.

[46] I am stating a general view of what counts as "enough" here, since my main concern is to defend a sufficientarian position on distributive justice over an egalitarian one. However, there is considerable room for dispute within particular contexts for what should count as enough. See, for instance, Carina Fourie and Annette Rid (eds.), *What Is Enough?: Sufficiency, Justice, and Health* (Oxford: Oxford University Press, 2017).

The specific means we take to ensure that people have enough will be a matter for prudential judgment, though it seems that it should involve some combination of law-governed markets, government assistance, and personal charity.

What I want to emphasize is that this sufficientarian conception of distributive justice, unlike luck egalitarianism, fits with a proper accepting-appreciating stance toward the given, since it first of all seeks to cherish, preserve, and foster the gift (or given good) of human life. Secondly, it also seeks to preserve the given good of the bond of the political community. Instead of engaging in "a metaphysical crusade against contingency," we should follow Wiggins' advice and:

> embark on the simpler and altogether more positive alternative project of engaging all citizens as fully as possible, in as many ways as possible, but each in the way best suited to their own aptitudes and predisposi-tions, with the shared thing in which they all participate, and of removing some of the greatest obstacles to this.[47]

Roger Scruton holds a similar view, as he maintains that a well-functioning political community depends upon a strong sense of "we," that is, a first-person plural, and he writes:

> No such first-person plural can emerge in a society divided against itself, in which local antagonisms and class war eclipse every under-standing of a shared destiny.... [Thus, we have to find] ways of spreading the benefit of social membership to those who have not succeeded in gaining it for themselves.

Scruton goes on to say: "the more we take from this arrangement, the more we must give in return. This is not a contractual obligation. It is an obligation of gratitude."[48] We might also call this an obligation of patri-otism, which again is analogous to filial piety, since we are recognizing an obligation to show gratitude to a source of our existence (or way of life).

---

[47] Wiggins, *Ethics*, 306.
[48] Roger Scruton, *How to Be a Conservative* (London: Bloomsbury, 2014), 41–2. On class warfare, see references in n. 9.

We should note—in light of what Scruton says about local antagonisms and class war eclipsing a sense of shared political destiny—that while a sufficientarian view of distributive justice does not hold that it is important that everyone has the same, this does not mean we should be unconcerned with growing economic inequality. This too can harm the bond of the political community, especially where wealth is allowed to influence politics and where it is strongly connected with social status. Any democratic government that is concerned not to become a plutocracy must take measures to avoid, as much as possible, allowing wealth to influence politics. The connection between wealth and social status, on the other hand, is, in part, a matter of culture: to lessen this connection we need to promote a culture in which what matters more than wealth is virtue and one's contribution to the common good. However, such a culture is hard to cultivate and maintain when there is great disparity in wealth, and where so much hangs on whether you make it into an elite sector of society. When people work hard to make ends meet, while others live in great abundance, it is hard for the former not to feel there is something unfair about such an arrangement. We need thus to find ways to spread the benefits of social membership more fairly in order to attain what I described earlier as a broad equality of condition, where everyone has enough to live well and on terms of equal respect with others.

Against Utopianism: The Politics of Imperfection

Sufficientarian justice is concerned with ensuring that every member of the political community has enough, and this implies a connection with the virtue of contentment, which I have described as the virtue of knowing when enough is enough, of not wanting more than is needed for a good life. When we consider the implications of this limiting virtue to the political domain, what it entails is an embrace of a politics of imperfection rather than a politics of perfection or utopianism.[49] A politics of imperfection does not mean quietism. In many ways we

---

[49] I have borrowed the term "politics of imperfection" from Anthony Quinton, *The Politics of Imperfection* (London: Faber and Faber, 1978).

ought to seek improvement, but a politics of imperfection requires us to acknowledge that perfection in politics is not feasible, and the attempt to achieve it often brings about greater problems than those that it sought to overcome (as we see in the utopian political projects of the twentieth century). It is significant that the word "utopia" is formed from the Greek words *ou* ("not") and *topos* ("place"), which suggests that a utopia is a "no place": it exists *nowhere*, except for in the imagination. A politics of imperfection is an embrace of *somewhere* with all its imperfection: we should seek improvement where needed, but we also need to find a way of being at home in the world amidst imperfection. The goal we should pursue is a *good enough* condition.

The issue of feasibility is often raised against egalitarian schemes of distributive justice. For instance, Nozick notes how freedom upsets equality: if people are allowed to be free with regard to the use of property, then, given their differences in natural talents and inclinations, they will arrive at unequal circumstances.[50] To try to achieve strict equality with respect to economic resources, for instance, will require us to sacrifice freedom. Luck egalitarians like Cohen do not require strict equality, since they allow that "genuine choice excuses otherwise unacceptable inequalities," but inequalities resulting from brute luck should be redressed. David Miller notes that there is still a significant feasibility problem here:

> To put [luck egalitarianism] fully into practice would require some agency capable of monitoring the situation of each individual person and working out how far their present resource level could be attributed to "luck" on the one hand and "choice" on the other, and then calculating, counterfactually, what their position *would* have been if the effects of luck had been neutralized and only choice remained. Then the agency would have to extract resources from the beneficiaries of "good luck" in order to provide compensation to the beneficiaries of "bad luck."[51]

---

[50] Nozick, *Anarchy, State, and Utopia*, 160–4.
[51] David Miller, *Justice for Earthlings: Essays in Political Philosophy* (Cambridge: Cambridge University Press, 2013), 2.

Miller notes that someone sympathetic to luck egalitarianism may respond that luck egalitarianism is "not supposed to guide practice directly" but rather it is supposed "to define justice at the most abstract level." It is then another question how far this can be implemented, and while there may be practical limitations to how far it can be applied, "these limitations should not influence the way we think about justice itself," since this would be to allow "our theory of justice to be contaminated by irrelevant contingencies." Such a response is that of someone who endorses an "ideal theory" of justice, as opposed to a "nonideal theory," which affirms that contingencies of human life are relevant for how we think about what justice requires.[52] Cohen, in fact, is a prominent proponent of ideal theory; he maintains that the principles of justice are "fact-insensitive" in not being subject to practical considerations of feasibility: "If justice is … each person getting her due, then justice is her due irrespective of the constraints that might make it impossible to give it to her."[53]

It is worth considering here Cohen's case for socialism in *Why Not Socialism?* He begins by presenting a context in which he thinks most people would favor socialist principles of equality and community, namely, a camping trip with friends, where everything is shared in common. The socialist principle of equality, he thinks, translates as the luck egalitarian position. However, Cohen says that the socialist principle of community would temper the inequalities allowed by luck egalitarianism (namely, those inequalities arising from choice rather than brute luck). This principle involves communal reciprocity where people care about each other and therefore seek to serve one another, and find satisfaction in doing so. Cohen maintains:

We cannot enjoy full community, you and I, if you make, and keep, say, ten times as much money as I do, because my life will then labor under challenges that you will never face, challenges that you could help me to cope with, but do not, because you keep your money.[54]

---

[52] For a good collection of essays on ideal (or utopian) theory vs. nonideal (or realist) theory, see Michael Weber and Kevin Vallier (eds.), *Political Utopias: Contemporary Debates* (Oxford: Oxford University Press, 2017).

[53] Cohen, *Rescuing Justice and Equality*, 232, 252–3.

[54] G. A. Cohen, *Why Not Socialism?* (Princeton, NJ: Princeton University Press, 2009), 35.

According to Cohen, market reciprocity differs from communal reciprocity because the market is characteristically driven by greed and fear, where others are seen as possible sources of enrichment or as threats to one's success, which, he says, "are horrible ways of seeing other people" (40). In capitalist societies, Cohen goes on to say, "a person typically does not care *fundamentally*, within market interaction, about how well or badly anyone other than himself fares. You cooperate with other people... because *you* seek gain.... [Mutuality] is only a by-product of an unmutual and fundamentally *non*reciprocating attitude" (44–5).

It is hard to imagine Cohen characterizing capitalism in a worse possible light (self-serving, motivated by greed and fear), and socialism in a better possible light (other-serving, motivated by care for others), than he does here. Of course, it is also possible to characterize socialism in particular and luck egalitarianism in general in a bad light, where one is seen as motivated by envy and resentment toward others who possess more than one does.[55] Indeed, when Cohen says we cannot enjoy full community with others if they make a lot more money than us, I do not see why this must be the case, unless we are motivated by envy and/or resentment. As I have argued, economic equality is not itself important; what matters is that everyone has enough for a good human life, and it is this sufficientarian view of distributive justice that is important for maintaining the bond of political community.

I think greed is a problem within capitalist economic orders, and I will return in Chapter 4 to discuss how we can counteract such greed. But I do not think economic incentives as such, where one is concerned to make a decent living, are morally problematic; indeed, any viable economic system, it seems, will have to depend upon them. To care about attaining through one's own efforts—and in cooperation with others— the material conditions for living a good life is a respect-worthy human endeavor, which also connects up with the dignity of human work (as I will discuss in Chapter 4). I also do not think human relationships within market interactions are as bleak as Cohen makes them out to be;

---

[55] Another possible unflattering interpretation is that one is motivated by a sense of moral superiority over the selfish capitalist (though, as Cohen himself notes, many professed egalitarians are, in fact, quite rich).

people can and do form friendships in market relationships where they wish well to each other for the friend's sake, and market relations also take place within broader bonds of patriotic loyalty and human solidarity. If things are really as bad as Cohen says, then it is not clear how it could be possible to summon anything near the kind of solidarity he needs for his socialist ideal.

The question of feasibility is the key problem that Cohen confronts with regard to his socialist ideal. He notes that the aspiration of socialists is "to realize the principles that structure life on the camping trip on a national, or even on an international, scale" (46). However, it is not clear that these principles will even work on the campgrounds. One will likely be willing to share freely among friends, but what about other campers not in one's group? What if there are moochers who seek to take food and supplies from others, even though they are capable of providing them for themselves? The problem of the moocher (or free-rider), especially when widespread, also seems corrosive to political community.

Cohen mentions two common challenges to realizing socialism on a national and international scale: first, it is often claimed that people are "by nature insufficiently generous and cooperative to meet its requirements"; second, even if they were sufficiently generous, or could become so, "we do not know how to harness that generosity; we do not know how...to make generosity turn the wheels of the economy," whereas we know how to harness selfishness through the market (55–6). Cohen thinks that it is the second problem that is the principal one: "Our problem is not, primarily, human selfishness, but our lack of a suitable organizational technology: our problem is a problem of *design*" (57–8). In thinking that the primary problem is one of proper design, Cohen finds himself in agreement with many other utopian thinkers. He acknowledges that the key advantage of capitalism is that market signals allow for efficiency in production (which, we should add, has done an enormous amount to reduce poverty around the world, even while also increasing economic inequality), and so the question arises of whether the socialist can make use of these market signals without relying on typical market incentives. This requires that people pursue the economic gain of markets for the sake of the community and not for themselves, which Cohen acknowledges is a utopian ideal (see 62–5). He goes on to write: "We socialists don't *now* know how to replicate camping trip procedures on a nationwide scale," or "how to give collective ownership and equality the real meaning it has in the camping trip story," or "how to honor personal choice, consistently with equality and community, on a large scale society" (75–6). However, he says: "we should never forget that greed and fear are repugnant motives" (77). Despite feasibility challenges to the socialist ideal, Cohen maintains that "they are not reasons to disparage the ideal itself" (80).

But we should question the ideal, and feasibility challenges are key reasons why. Cohen is wrong to think that the design challenge is the main problem; rather, the limits of human sympathy and generosity are the more important reason why we will never realize the socialist utopia. Indeed, as Hume contends, it is because of the limits of sympathy and benevolence and the moderate scarcity of material resources that we need justice, which he describes as a "cautious, jealous virtue."[56] We can go further: we need justice not only because of limited sympathy but also because of human tendencies toward rapaciousness, dominance, cruelty, etc. We need to acknowledge that human beings are a mixed bag; to use Hume's language, while we may have a "particle of the dove" within us (i.e., tendencies toward sympathy and benevolence, albeit limited), this exists alongside "elements of the wolf and serpent."[57]

Hume, it should be said, overstates things when he describes justice as a *jealous* virtue. Justice is not jealous, but rather it is concerned with giving everyone his or her due, which includes avoiding overreaching for benefits, where one goes beyond what one deserves and does so at the expense of others, which is the vice of *pleonexia*, as Aristotle describes it (NE V.2). Aristotle in fact says something similar to Hume when he remarks: "if people are friends, there is no need for justice" (NE VIII.1, 25–6). Aristotle does think there should be a kind of friendship among

---

[56] David Hume, *An Enquiry Concerning the Principles of Morals* (1751), in *Enquiries Concerning Human Understanding and Concerning the Principles of Morals*, 3rd edn (Oxford: Clarendon Press, 1975), 184. Hume gives a similar account in his *A Treatise of Human Nature*, where he says: "if men were supplied with every thing in the same abundance, or if *every one* had the same affection and tender regard for *every one* as for himself; justice and injustice would be equally unknown among mankind" (*A Treatise of Human Nature*, 2nd edn [Oxford: Clarendon Press, 1978 (1739)], 495).

[57] Hume, *An Enquiry Concerning the Principles of Morals*, 271.

citizens of the *polis*, but this is limited, and so justice is needed as well. Indeed, as Wiggins puts it in his "Neo-Aristotelian Reflections on Justice," the *polis* (or political community) is "an association for sharing or *koinōnia*" that enables us to live and live well, and "justice is that by which this *koinōnia* is maintained."[58] Certainly this will require rectifying wrongs done, but we have seen Wiggins maintain that it is especially dire need that justice should address, since this poses a major threat to the social bond.

Cohen also maintains that justice is giving everyone his or her due, though this is understood according to a socialist ideal that he acknowledges may not be feasible. We have seen that Cohen thinks that feasibility challenges "are not reasons to disparage the ideal itself," since justice is due "irrespective of the constraints that might make it impossible" to realize it. Cohen explicitly regards his position as a "Platonic" one, which seeks to grasp "justice in its purity," that is, in its ideal form.[59] However, we can contrast Cohen's "Platonic" conception of justice with an "Aristotelian" conception that acknowledges that justice is constrained by contingent reality, and in particular the contingent reality of the human condition, since justice should be concerned with enabling us to live and live well in human community. If we take the Aristotelian conception, as I think we should, then feasibility challenges do give us reasons to question utopian schemes such as Cohen's. A key difference here is over what we think is the prime concern for a conception of justice: is it to try to understand the ideal form of society, or is it to guide social practice in the human condition as we have it? The "Aristotelian" view maintains the latter position, whereas the "Platonic" view maintains the former. Cohen expresses the Platonic view when he writes: "One may or may not care about practice, but one may also care about justice, as such, one may be interested in what it is, even if one does not care about practice at all."[60] However, it is strange to think that one could care about

---

[58] Wiggins, *Ethics*, 281; see also 312.   [59] Cohen, *Rescuing Justice and Equality*, 291.
[60] Cohen, *Rescuing Justice and Equality*, 306. Cohen goes on to say:

> It would, of course, be unusual to be interested in what justice is while not caring at all about practice. Being interested in what justice is standardly goes with caring about it, and, therefore, with caring about whether practice is appropriately responsive to it.... [But] I want to know what justice is whatever I or anyone else may think is the right form and amount of the contribution that justice should make to political

what justice is without caring about practice. Justice should be understood as practice-guiding; otherwise, it is not clear why we should care about justice at all.

Now one could argue that utopian schemes are important for guiding practice precisely by giving us an ideal to aim at in our practice, even if we can never fully attain the ideal. I do not deny the importance of ideals in the sense of aspirations for making life better, but it is when they are utopian that they are highly problematic and should be rejected. My fundamental complaint against utopianism is the same as against luck egalitarianism, but it applies to an even greater extent, namely, utopianism wages war against the given world, and thereby fails to address our need for being at home in the given world, which is a need that Cohen himself acknowledges, as we saw in Chapter 1 when considering his account of accepting the given.

This point about utopianism waging war against the given world is brought out in several important critiques of utopianism. For instance, Aurel Kolnai writes that "utopias are in fact intrinsically at war with the basic structure of reality," and thus, strictly speaking, they are unrealizable; nevertheless, they offer "'another world' *in place of the world*": "The purity, exclusiveness and isolation of the utopian dream-world proclaim the negation of the world, i.e. of the world of division, of mutual awareness of alterity, with the ever-renewed demands of adjustment, of friction, compacts, and compromise."[61] The basic structure of reality that utopianism is at war with is not only the inequalities of natural endowments, our limited sympathy, and the mixed bagness of human nature

and social practice. I personally happen also to be exercised by the latter question, but one need not be exercised by it in order to care about the first one.   (307)

[61] Aurel Kolnai, "The Utopian Mind," in *Privilege and Liberty and Other Essays in Political Philosophy* (Lanham, MD: Lexington, 1999), 122, 124. In a book of the same title, Kolnai writes:

Utopianism rejects the commonsense *submission* to the human condition and "pursuit of the good" *on its terms*, erecting in its place the idol of a perfect human condition which would not be a human condition and which demands the *self-surrender* of man to an alien, unreal and (as to its actual features) inconceivable construct of his abstractive mind. Linked therewith is a mirage of the all-comprehensive perfect good, discontinuous and out of tune with man's actual pursuit of value achievements in the framework of a reality which is not identical with the Good but logically inseparable from any meaning man may attribute to it.
(*The Utopian Mind and Other Papers*, ed. Francis Dunlap
[London: Athlone, 1995], 101)

but also the fact that there are many good things we seek to attain that cannot all be fully realized together (Cohen seems to recognize this when he speaks of the tensions between freedom, equality, and community), while at the same time there are many bad things that we are seeking to avoid that cannot all be avoided. And so we will have to make choices about how best to bring out the good and avoid the bad, knowing that perfection cannot be achieved, and we must do so in a social world where other agents are seeking to do the same, but where they do so in ways that can come into conflict with our aims, and so compromise is often necessary (here is another reason why utopia will never be realized: people will not agree on the destination). However, Roger Scruton contends that the unrealizability of utopian visions is, in fact, part of their appeal:

> Those who espouse them are aware that final solutions are not available, that conflicts and competition are essential features of human societies, and that all attempts to achieve a permanent unity of purpose or absolute equality of condition are incompatible with the freedoms required by peaceful coexistence among strangers. Utopians must therefore live in a condition of constant preparation, fighting the enemies of utopia, and knowing that the fight will never end. In other words, utopia exists as a great negation sign, *un gran rifiuto*, to be affixed to everything actual and to command and authorize every form of violence against it . . . . It serves . . . as an abstract condemnation of everything around us, and it justifies the believer in taking full control.[62]

We have seen that in the twentieth century utopian schemes, in fact, have been used to justify terrible violence. Cohen himself does not attempt to justify violence, but his utopianism does serve to foster disappointment with the given world. As David Miller puts it, Cohen's political philosophy functions as a kind of lamentation, given the distance that separates

---

[62] Roger Scruton, *The Uses of Pessimism: And the Dangers of False Hope* (Oxford: Oxford University Press, 2010), 69–70.

his ideal social world and the world as it exists, and given the infeasibility of the ideal.[63]

We have seen—in Chapter 1—that Cohen endorses the importance of the limiting virtue of contentment, where we see the glass as half-full, where we look for the good in everything, and believe what is good is often good enough and that it is often wiser to satisfice than to maximize, and we have seen how this is important for being at home in the world, or as Cohen puts it, for being "at peace with the world" and therefore with ourselves. The problem is that Cohen does not allow these important reflections to constrain his conception of justice (he does allow that we may have to balance the claims of justice against other values, but this is not the same as constraining our conception of justice). Each of the limiting virtues, in fact, needs to be understood as constraining our conception of justice. In the next section I will discuss the virtue of moderation, but here I will briefly indicate how other limiting virtues besides moderation and contentment should constrain our conception of justice.

As discussed in the first section of this chapter, what is due to others partly depends on the extent to which we stand in a relationship of proximity, where the virtue of neighborliness applies; it also depends upon whether we stand in a relationship of proper loyalty, such as to family, friends, and fellow citizens. Justice should also be understood as constrained by the virtue of reverence, which calls on us to be properly responsive to the special dignity or sanctity of human life and not to transgress against it (recall the discussion of absolute prohibitions in Chapter 2). But for the sake of seeing what is wrong with utopianism, the limiting virtue that is especially important to recognize, along with the virtues of moderation and contentment, is the virtue of humility, which, as discussed in Chapter 1, is concerned with reining in our tendency to "play God." Utopian schemes often give expression to a Promethean mindset that seeks bring about heaven on earth (or build a tower to heaven). However, the virtue of humility requires us to recognize our inability fully to understand and control complex human affairs and direct them to our ideal ends. Indeed, as informed by the light of human history, it

---

[63] See David Miller, "A Tale of Two Cities; or, Political Philosophy as Lamentation," in *Justice for Earthlings*, 228–49.

requires that we acknowledge that the best-laid schemes of human beings regarding complex human affairs often go awry, and that in our attempts to bring about heaven on earth we can in fact bring about hell.

It is better to focus on realizing the good that is possible, and one of the most important reasons for objecting to utopian thinking is that it is counterproductive to realizing this good. Rather than imagining some ideal, yet unrealizable world by which we critique the actual world, we do better to focus on becoming better people ourselves. As Max Scheler puts it: "Repentance, not Utopia, is the greatest revolutionary force in the moral world."[64] We also do better to seek to give our appreciative attention to what is good in the given world, and what is tried-and-true, which can also help us to realize the good more fully. As Edmund Burke says in his reflections on the French Revolution:

> [In] general, those who are habitually employed in finding and display-ing faults, are unqualified for the work of reformation: because their minds are not only unfurnished with patterns of the fair and good, but by habit they come to take no delight in the contemplation of those things. By hating vices too much, they come to love men too little.... From hence arises the complexional disposition of some ... to pull every thing in pieces.[65]

Finally, utopian thinking is counterproductive for the work of prudential compromise that is needed to solve human conflicts, where we must look at each case of conflict on its own terms. Here prudence will be connected with the limiting virtue of moderation to which I now turn.

Moderation and the Limits of Government

As discussed in Chapter 2, the virtue of moderation is concerned with avoiding vicious extremes. Regarding the political domain, it must be

---

[64] Quoted in Kolnai, "The Utopian Mind," 131.
[65] Edmund Burke, *Reflections on the Revolution in France* (1790), in *Select Works of Edmund Burke, Volume 2* (Indianapolis, IN: Liberty Fund, 1999), 277.

said that we live in what is in many respects an age of extremes, as political polarization and fragmentation abound, which leads to increased conflict and threatens the bond of political community. Moderation is important for counteracting this polarization and fragmentation and thus for preserving the bond of the political community and for realizing the good that is possible in such community.

But what exactly does moderation mean in the political domain? Drawing on Aurelian Craiutu's work,[66] I suggest the following are important aspects of political moderation: Political moderates embrace a politics of imperfection and acknowledge that conflicts will always be with us, and so this means that we must find paths of compromise. This does not mean that there are not some things on which we should not compromise; as we have seen in Chapter 2, Aristotle—the great advocate of moderation—acknowledges that there are actions for which there is not a proper mean. There are some grave injustices that we should not accept, such as murder, torture, and slavery (unfortunately Aristotle himself endorsed the last). What political moderates are committed to is an ethic of dialogue and civility: they seek to communicate across differences of view in order to find common ground; they seek to do so respectfully, charitably, and with an open mind; and they see promoting and embodying this ethic of dialogue and civility as a way of preserving the bond of political community in the face of continued disagreement.

Political moderates seek to avoid ideological Manichaeism where one's side is seen as a force of light and goodness, and the other side is seen as a force of darkness and evil. The political moderate will recognize that there is likely to be some element of truth on different sides of political debates, and they will also recognize conservatism and progressivism as dispositions rooted in human nature: in so far as there are good things worthy of love and attachment, we will want to conserve them; in so far as there are bad things, we will want change for the better. The challenging task of political prudence and moderation is how best to bring out the good and avoid the bad in a situation where people often make different judgments about the nature and extent of the good and the bad. Political

---

[66] Aurelian Craiutu, *Faces of Moderation: The Art of Balance in an Age of Extremes* (Philadelphia, PA: University of Pennsylvania Press, 2017).

moderates oppose ideological purity tests and thus are nonconformists (which can at times require courage to stand against groupthink). They also oppose radicalism, fanaticism, dogmatism, extreme partisanship, and the like. Political moderates prefer gradual, piecemeal reform to revolution and sweeping change; they seek changes on which common ground can be found and that are necessary for keeping the political community in a state of equipoise with regard to various contending factions. This does not mean that there might not be some circumstances in which sweeping changes are needed because things are so bad, but the political moderate is resistant to the tendency to find political emergencies everywhere. Political moderates are concerned above all to avoid the worst evils or bad things—such as unsatisfied dire need, cruelty, violence, and anarchy—rather than to achieve the most ideal state of affairs. Political moderation, Craiutu writes, "is a humble expression of imperfection and limitations" (236).

Craiutu also notes: "Political moderation has traditionally been linked to the separation and balance of powers, the use of the executive veto, bicameralism, neutral power (or third power) as well as the adoption of declarations or bills of rights" (230). We can see these features in the US Constitution: there is a separation of powers between three branches of government—executive, legislative, and judicial—which provide "checks and balances" to each other; there is executive veto with the president; there is bicameralism with the Senate and the House of Representatives; the judiciary is supposed to be a neutral power; and there is a bill of rights. The authors of *The Federalist* in fact sought to argue for the ratification of the Constitution through appeal to the wisdom of political moderation gained from history and human experience.[67] In *Federalist* No. 1 Alexander Hamilton speaks of the role in human affairs of "[ambition], avarice, personal animosity, party opposition, [etc.]" as "inducements to moderation," and he is concerned to balance the efficiency of the federal government and the union between the states with liberty. In *Federalist* No. 6 he says that we should reject those "Utopian

---

[67] Alexander Hamilton, John Jay, and James Madison, *The Federalist*, Gideon edition (Indianapolis, IN: Liberty Fund, 2001 [1787–8]), 43. See also Peter Berkowitz, *Constitutional Conservatism: Liberty, Self-Government, and Political Moderation* (Stanford, CA: Hoover Institution Press, 2013), ch. 3.

speculations" and "idle theories" that promise us "an exemption from the imperfections, the weaknesses, and the evils incident to society in every shape," and instead we should "adopt as a practical maxim for the direction of our political conduct, that we, as well as the other inhabitants of the globe, are yet remote from the happy empire of perfect wisdom and perfect virtue." In *Federalist* No. 9 Hamilton identifies the "regular distribution of power into distinct departments; the introduction of legislative balances and checks; the institution of courts composed of judges, holding their offices during good behaviour; the representation of the people in the legislature, by deputies of their own election," as the "means, and powerful means, by which the excellencies of republican government may be retained, and its imperfections lessened or avoided."

In *Federalist* No. 10 James Madison argues that a well-constructed constitution—which he believes the US Constitution to be—should above all be concerned with controlling the effects of faction, which he understands as "a number of citizens, whether amounting to a majority or minority of the whole, who are united and actuated by some common impulse of passion, or of interest, adverse to the rights of other citizens, or to the permanent and aggregate interests of the community." The goal is to control the effects of faction rather than to remove it altogether because that would mean abolishing liberty.[68] In *Federalist* No. 51 Madison writes:

> It may be a reflection on human nature, that such devices [as the separation of powers and the system of checks and balances] should be necessary to control the abuses of government. But what is government itself, but the greatest of all reflections on human nature? If men were angels, no government would be necessary. If angels were to govern men, neither external nor internal controls on government

---

[68] Madison writes:

> Liberty is to faction, what air is to fire, an aliment, without which it instantly expires. But it could not be a less folly to abolish liberty, which is essential to political life, because it nourishes faction, than it would be to wish the annihilation of air, which is essential to animal life, because it imparts to fire its destructive agency.
>
> (*Federalist* No. 10)

would be necessary. In framing a government which is to be administered by men over men, the great difficulty lies in this: you must first enable the government to control the governed; and in the next place oblige it to control itself. A dependence on the people is, no doubt, the primary control on the government; but experience has taught mankind the necessity of auxiliary precautions.

Madison is not saying that human beings are wholly vicious; indeed, in *Federalist* No. 55 he notes that republican government requires virtue to a significant degree. But we cannot expect people to be perfect in virtue, and so government should seek to counteract human imperfection, including as it operates within the government itself. The separation of powers and the system of checks and balances, therefore, express a form of limited government, as the powers of government are thereby limited. The US Constitution also provides a limited form of government in that, as Madison puts it in *Federalist* No. 45, "[the] powers delegated … to the federal government are few and defined," and all other powers, as the Tenth Amendment puts it, "are reserved to the States respectively, or to the people."

Political moderation commits us to limited government, that is, it commits us to putting the brakes on the tendency toward unlimited (which is to say, totalitarian) government. It is important not to expect government to do too much. Many of the most important pursuits in life—such as our pursuits with respect to love, friendship, family life, religious devotion, intellectual enquiry, meaningful work, projects of self-cultivation, etc.—lie outside of the political domain, and politics should make space for these pursuits. It is especially important that we do not seek our salvation or ultimate meaning through politics, which is the error of utopian schemes (I do not deny, of course, that we can legitimately find *some* meaning in life through political engagement). On this point, consider Albert Camus:

Politics must, wherever possible, be put back in its rightful place, which is a secondary one. Its aim should not be to provide the world with a gospel, or a catechism … The great misfortune of our time is precisely that politics pretends to furnish us … with a catechism, a complete philosophy... But the role of politics is to set our house in order, not to deal with our inner problems. I do not know for myself whether or not there is an absolute. But I do know that this is not a political concern. The absolute is not the concern of all: it is the concern of each.[69]

One way that politics can seek to offer a gospel, or a catechism, is through a progressive narrative that provides a path for realizing a kind of salvation from the ills of the human condition, which can threaten despair. In *The True and Only Heaven: Progress and Its Critics*, Christopher Lasch explores how a prominent form of progressive optimism, which believes in a "march of history" toward some ideal or ever-better state of affairs, operates as a secular religion. He begins by asking: "How does it happen that serious people continue to believe in progress [in the march of history sense], in the face of massive evidence that might have been expected to refute the idea of progress once and for all?"[70] Lasch thinks that belief in progress has to be understood as "an antidote to despair," and with the calamities of the twentieth century and the collapse of grand utopian visions it has become clear to many that "a belief in progress could be salvaged... only by disavowing its perfectionist overtones" (41). He writes:

The concept of progress can be defended against intelligent criticism only by postulating an indefinite expansion of desires, a steady rise in the general standards of comfort, and the incorporation of the masses into the culture of abundance. It is only in this form that the idea of progress has survived the rigors of the twentieth century. More extravagant versions of the progressive faith, premised on the perfectibility of human nature—on the unrealized power of reason or love—collapsed a long time ago; but the liberal version has proved surprisingly resistant to the shocks to easy optimism administered in a rapid succession by

---

[69] Albert Camus, "The Human Crisis" (1946), trans. Lionel Abel, in *Civil Liberties and the Arts: Selections from* Twice a Year, *1938–48*, ed. William Wasserstrom (Syracuse, NY: Syracuse University Press, 1964), 250. See also Jean Bethke Elshtain, *Augustine and the Limits of Politics* (Notre Dame, IN: University of Notre Dame Press, 1995); Elshtain, *Sovereignty: God, State, and Self* (New York: Basic Books, 2008).

[70] Christopher Lasch, *The True and Only Heaven: Progress and Its Critics* (New York: Norton, 1991), 13.

twentieth-century events. Liberalism was never utopian, unless the democratization of consumption is itself a utopian ideal. It made no difficult demands on human nature. It presupposed nothing more strenuous in the way of motivation that intelligent self-interest.   (78)

At the end of his book, Lasch remarks: "Progressive optimism rests, at bottom, on a denial of the natural limits on human power and freedom, and it cannot survive for very long in a world in which an awareness of those limits has become inescapable" (530). Lasch thinks that in order to face properly the challenges of our age, as well as the inescapable tragic dimension of human life, we need to cultivate a proper "sense of limits," or "respect for limits," which means cultivating a disposition that he describes variously as hope, trust, or wonder, which "asserts the goodness of life in the face of its limits," and which involves a "grateful acceptance of a world that was not made solely for human enjoyment" (15, 17, 530; see also 46–7, 80–1).

I want to turn now to examine the capitalist form of progressive optimism that postulates "an indefinite expansion of desires, a steady rise in the general standards of comfort, and the incorporation of the masses into the culture of abundance," and I will also consider a particular utopian vision of capitalism. Against such views, I will make the case for a respect for limits in the economic domain as important for the good life.

4

Economic Limits

In making the case for economic limits I will begin by considering the problem of greed in modern economic life as the key problem that needs to be addressed. I will discuss the importance of the limiting virtue of contentment here for counteracting greed. In the second section I will develop a positive vision for the role of economics in the good life. I call this "home economics," as it seeks to recover something of Aristotle's idea of *oikonomia* as centering on the home (*oikos*), but also contributing to the common good of the particular communities to which one belongs. What we need, I contend, is a way of living our economic life that contributes to our being properly at home in the world, rather than causing alienation (or not-at-home-ness). This will also require recovering a sense of the dignity of work. In the third and final section I will discuss the practice of the Sabbath (as a specific form of leisure) as providing an important limitation on work, while also in a crucial sense completing our work, and I will suggest its relevance for everyone, regardless of religious beliefs. This discussion will also provide a fitting conclusion to the book, since it will return us to the discussion with which the book began: the two fundamental existential stances—the choosing-controlling stance and the accepting-appreciating stance—and the proper relationship between them.

The Vice of Greed and the Virtue of Contentment in Economic Life

Our capitalist economic order seems to be a "two-edged sword," write father-son economist-philosopher duo Robert and Edward Skidelsky: "On the one hand, it has made possible vast improvements in material conditions. On the other, it has exalted some of the most reviled human

The Virtues of Limits. David McPherson. Oxford University Press. © David McPherson 2022.
DOI: 10.1093/oso/9780192848536.003.0005

characteristics, such as greed, envy and avarice."[1] It is not entirely clear how the Skidelskys are distinguishing greed and avarice here; it seems that they are using greed as a broader term that denotes a selfish acquisitive desire that is insatiable, and avarice is a specific form of this where the desire is for the acquisition of wealth. However, I will use greed as synonymous with avarice, since this is a common usage, and greed is the more familiar term. (I will also use money and wealth as basically synonymous, even though wealth can be regarded as a broader term that includes assets, such as real estate and other property, which can be translated into money.) The Skidelskys mention that greed is one of the most reviled human characteristics, but we should add a qualification to this statement, and say that at least traditionally it has been. In our capitalist economic order, it is now an open question whether greed should be shunned. We can frame the question here as: is there anything wrong with wanting more and more wealth?

One answer to this question is to say that while greed may be a personal vice, nevertheless it should be accepted as conducive to public benefit. Such was the position of Bernard Mandeville in *The Fable of the Bees: or, Private Vices, Public Benefits*:

> The Root of Evil, Avarice,
>
> That damn'd ill-natur'd baneful Vice,
>
> Was Slave to Prodigality
>
> That Noble Sin; whilst Luxury
>
> Employ'd a Million of the Poor,
>
> And odious Pride a Million more:
>
> Envy it self, and Vanity,
>
> Were Ministers of Industry.[2]

Adam Smith takes a similar position (indeed, he was influenced by Mandeville), though he has greater ambivalence about greed, and he emphasizes the importance of cultivating virtue for maintaining social

---

[1] Robert Skidelsky and Edward Skidelsky, *How Much Is Enough?: Money and the Good Life* (New York: Other Press, 2012), 3.

[2] Bernard Mandeville, *The Fable of the Bees: or, Private Vices, Public Benefits* (1714), in *The Fable of the Bees and Other Writings* (Indianapolis, IN: Hackett, 1997), 28.

order and addressing individual corruption.[3] In his first discussion of the "invisible hand" in *The Theory of Moral Sentiments*, Smith does come close to Mandeville in explicitly endorsing greed because of its public benefits, writing:

[In] spite of their natural selfishness and rapacity, though they mean only their own conveniency, though the sole end which they propose from the labours of all the thousands whom they employ be the gratification of their own vain and insatiable desires, [the rich] divide with the poor the produce of all their improvements. They are led by an invisible hand to … advance the interest of the society.[4]

However, in his later discussion of the invisible hand in *The Wealth of Nations*, Smith switches to talking about self-interest (or self-love):

[Every] individual [in his industry] … neither intends to promote the public interest, nor knows how much he is promoting it…. [By] directing that industry in such a manner as its produce may be of the greatest value, he intends only his own gain, and he is in this, as in many other cases, led by an invisible hand to promote an end which was no part of his intention…. By pursuing his own interest he frequently promotes that of the society more effectually than when he really intends to promote it.[5]

Indeed, he famously says:

It is not from the benevolence of the butcher, the brewer, or the baker, that we expect our dinner, but from their regard to their own interest. We address ourselves, not to their humanity but to their self-love, and never talk to them of our own necessities but of their advantages.[6]

---

[3] For a good discussion of these matters, see Ryan Patrick Hanley, *Adam Smith and the Character of Virtue* (Cambridge: Cambridge University Press, 2009).

[4] Adam Smith, *The Theory of Moral Sentiments* (Indianapolis, IN: Liberty Fund, 1985 [1759]), pt. IV, ch. 1.

[5] Adam Smith, *An Inquiry into the Nature and Causes of the Wealth of Nations*, vol. I (Indianapolis, IN: Liberty Fund, 2009 [1776]), bk. IV, ch. II.

[6] Smith, *The Wealth of Nations*, bk I, ch. II.

As mentioned in Chapter 3 when responding to Cohen's argument for socialism, I do not think there is anything wrong with economic incentives as such; indeed, any viable economic system, it seems, will depend upon them, and they are often connected to a desire to secure for oneself the material conditions of a good human life. Greed, on the other hand, is a vice to be shunned, and there still seems to be an endorsement of greed in Smith's later discussion of the invisible hand when he speaks of someone directing his industry so that "its produce may be of the greatest value." However, before starting to make the case against greed, I want to consider one more prominent defense of greed that appeals to its social benefits.

I have in mind John Maynard Keynes, who sought to justify greed within capitalism as a necessary evil for achieving a future utopia of widespread abundance with very little work, and thus overcoming the curse of Adam, that is, the curse of having to toil for our sustenance (Genesis 3:17–19). In his 1930 essay "Economic Possibilities for Our Grandchildren," Keynes maintains that in a hundred years—thanks to continued economic growth and advancing technology—we will have solved what he calls the "economic problem," that is, the "struggle for subsistence," and people will work far less and earn far more: "for the first time since his creation man will be faced with his real, his permanent problem—how to use his freedom from pressing economic cares, how to occupy the leisure, which science and compound interest will have won for him, to live wisely and agreeably and well."[7] At most, he thinks, we will have to work around fifteen hours a week. We will then be able to return to traditional virtue and affirm "that avarice is a vice, that the exaction of usury is a misdemeanour, and the love of money is detestable." However, he warns that we are not there yet:

> For at least another hundred years we must pretend to ourselves and to every one that fair is foul and foul is fair; for foul is useful and fair is not. Avarice and usury and precaution must be our gods for a little

---

[7] John Maynard Keynes, "Economic Possibilities for Our Grandchildren" (1930), in *The Collected Writings of John Maynard Keynes*, vol. IX: *Essays in Persuasion* (Cambridge: Cambridge University Press, 2015), 326, 328.

longer still. For only they can lead us out of the tunnel of economic necessity into daylight.   (331)

Keynes in fact describes the love of money as "a somewhat disgusting morbidity, one of those semi-criminal, semi-pathological propensities" (329), but he thinks capitalism depends on this, and so what he is suggesting is that we must live with acquisitive vice and pathology in the hopes that our grandchildren might one day be able to have virtue and extensive leisure.

The sacrificing of virtue for economic gain is, as the Skidelskys put it, a "Faustian bargain," a classic "deal with the devil" in which we "sell our soul" in return for some supposed good. However, deals with the devil never turn out well. The Skidelskys note that while Keynes was largely right about continued economic growth, he was wrong about the predicted reduction in work hours.[8] Why is this? One explanation they offer—in the vein of Marx's account of worker exploitation in capitalism—has to do with the way that free-market economies have allowed employers to dictate the terms of employment, which has led to increased inequality: those at the top have gained significantly from increased productivity, while ordinary workers have not realized gains in real income that would make it possible and desirable to work less.[9] Second, the Skidelskys maintain that we have an insatiable tendency to acquisitiveness that is driven by a concern for comparative social status, which capitalism exacerbates and which traditional forms of ethics have sought to curb (recall the discussion in Chapter 2 of Aristotelian and Confucian forms of character formation and how they seek to restrain our tendency to appropriate whatever we want in the world for own use and satisfaction) (7, 33–42). The Skidelskys identify four ways that capitalism exacerbates our insatiable desire for wealth (40–1): First, "capitalism's competitive logic drives firms to carve out new markets by (among other things) manipulating wants," namely,

---

[8] While the average hours of work have fallen to some degree, the decrease in work hours is nowhere near what Keynes predicted: "In 1930 people in the industrial world worked roughly a fifty-hour week. Today they work a forty-hour week" (Skidelsky and Skidelsky, *How Much Is Enough?*, 21). The Skidelskys also note: "While overall working hours have stalled, many lower paid workers are working less than they want to, while many of the rich are working more than they need to" (23).

[9] Skidelsky and Skidelsky, *How Much Is Enough?*, 7, 30–3.

through advertising. Second, "capitalism greatly broadens the scope of status competition," especially in connection with a democratic ethos: the "combination of social equality and income inequality has…become the capitalist norm, leading to a situation in which every member of society is in a sense competing against every other," and where the inequality is greater so too is the competitive pressure. Third, "the ideology of free-market capitalism has been consistently hostile to the idea that a certain sum of money could represent 'enough.'" Fourth and finally, capitalism increases the range of things for which money is a factor, and thereby it makes money matter more in human life.[10] The Skidelskys conclude: "Keynes's mistake was to believe that the love of gain released by capitalism could be sated with abundance, leaving people free to enjoy its fruits in civilized living…. Capitalism has achieved incomparable progress in the creation of wealth, but has left us incapable of putting that wealth to civilized use," namely, in leisure, which is where we do activities for their own sake and not under any compulsion (41–2).

In order to put wealth to civilized use the Skidelskys maintain that we need an objective conception of the good life (which makes claims on how we ought to live, regardless of what we may happen to desire); this would be "a life sufficient unto itself" (7), and thus it is in opposition to the doctrine of endless economic growth, which is senseless without some defined sense of what it is *for*. Here they are operating within the Aristotelian ethical tradition. When Aristotle explores what kind of life *eudaimonia* (human fulfillment, the good life, "living well and doing well") consists in, he briefly considers that it might consist in the money-making life, but he sets this aside, saying: "wealth is clearly not the good we are looking for, since it is useful and for the sake of something else" (NE I.5, 1096a5–7). While people speak of the love of money—we can think of Scrooge here—what is loved is, in fact, not money itself but rather the way in which it is a marker of social status and/or how it is useful for attaining other things we want; for instance, money can be used in service of the pleasure-seeking life. Regarding social status, Aristotle recognizes the importance of honor, but he thinks it is more

---

[10] This point is also explored in Michael J. Sandel, *What Money Can't Buy: The Moral Limits of Markets* (New York: Farrar, Straus and Giroux, 2012).

important to be *honorable* through being virtuous (I.5, 1095b21–9). Aristotle also does not think *eudaimonia* consists in the pleasure-seeking life, since it is "slavish" and a life "characteristic of grazing cattle" (I.5, 1095b18–19). Nevertheless, he thinks wealth is important for the virtuous life within human community, since "it is impossible or not easy to do noble actions without supplies" (I.8, 1099a30–1), and he thinks that *eudaimonia* consists in part in such noble or virtuous actions. We can think particularly of generosity as a virtue that is facilitated by wealth (IV.1–2). Aristotle also thinks that the just person will avoid the vice of *pleonexia*, which, as mentioned in Chapter 3, involves overreaching in benefits for ourselves at the expense of others (V.2)

Aristotle believes that *eudaimonia* above all consists in contemplation (see X.7–8), and he thinks wealth is also important for making possible the leisure wherein contemplation takes place; he writes: "happiness seems to reside in leisure, since we do unleisured things in order to be at leisure" (X.7, 1177b3–4). I think Aristotle has an overly negative view of unleisured work, which is also shared by Keynes, as we have seen, and to some extent by the Skidelskys in so far as they accept the terms of Keynes's discussion.[11] In the next section I will return to discuss the way in which work is a constitutive component of the good life, especially in regard to making a home in the world. What Aristotle, Keynes, and the Skidelskys overlook is how necessity is the mother not only of invention but also of virtue.[12] But they are right about the importance of leisure for the good life, and I will explore this further in the final section. I will just say for now that recognizing a place for leisure in the good life puts limits on the pursuit of wealth.

Let us return to the Skidelskys' argument. For the Skidelskys, it is an objective conception of the good life that enables us to set limits on acquisitiveness by specifying how much money is *enough*: we need enough to live a good life, and too much or too little can be harmful for our ability to realize such a life. They think a key reason why the

---

[11] See Skidelsky and Skidelsky, *How Much Is Enough?*, 27–33. For a helpful discussion of Aristotle's views on work, see Tom Angier, "Aristotle on Work," *Revue Internationale de Philosophie* 278:4 (2016): 435–49.

[12] Leon Kass makes this claim about necessity in *The Hungry Soul*, 107, 229. However, his focus is on eating and the virtues related to it.

insatiable desire for wealth has taken root and become so prominent in liberal societies is because there has been an eclipse of an objective conception of the good life: political liberalism—of the sort defended by John Rawls—seeks to be neutral between competing conceptions of the good life and instead provides people with a framework of rights and liberties by which they can pursue their particular conception of the good life (whatever it is) so long as they do not infringe upon the ability of others to do likewise.[13] However, liberalism is in fact not really neutral, since it ends up endorsing a particular conception of the good life, namely, the life of the choosing, preference-satisfying self which is not "encumbered" by any strong social ties and an objective conception of the good life. Against this, the Skidelskys make their case for an objective conception of the good life that can define the proper role of money in our lives. They put forward an account of basic goods that they believe are constitutive of the good life, namely, (1) health; (2) security; (3) respect; (4) personality (i.e., "the ability to frame and execute a plan of life reflective of one's taste, temperament and conception of the good," where this understood in objective terms: "choice *responds* to value"); (5) harmony with nature; (6) friendship; and (7) leisure.[14]

This is a plausible account of basic human goods (harmony with nature is perhaps the most contestable, though I would accept it where it is understood as part of being at home in the world), and I agree with the overarching argument that we need an objective conception of the good life to determine how much is enough with regard to wealth. However, it is noteworthy that the Skidelskys do not list and discuss the virtues here as basic goods that are constitutive of the good life, as is Aristotle's view and my own (virtue is its own reward because virtue is inherently noble and therefore fulfilling).[15] Without this there is a danger that the Skidelskys' account of basic goods may still leave too much room for acquisitiveness, especially given that it seems that a lot could fall under what they describe as the basic good of "personality," since this includes one's taste and temperament, even though they rightly say that

---

[13] See Skidelsky and Skidelsky, *How Much Is Enough?*, 86–7.
[14] Skidelsky and Skidelsky, *How Much Is Enough?*, ch. 6.
[15] See McPherson, *Virtue and Meaning*, ch. 2.

choice needs to be responsive to intrinsic values. The virtues, on my view, are modes of proper responsiveness to what is of intrinsic value (e.g., human dignity, or the nobility of virtue itself in realizing what is admirable in our humanity), and in being properly responsive they constitute for us the good life understood as a normatively higher, nobler, more fulfilling mode of life. The virtues also help to define for us how much wealth is enough; for instance, the virtues of generosity (or charity) and justice certainly help to answer the question "How much is enough?"

The virtue that is directly concerned with the question is, of course, the limiting virtue of contentment, which the Skidelskys do not explicitly explore in their book (though it is implicit in their topic). As I have defined it, contentment is the virtue of knowing when enough is enough, of not wanting more than is needed for the good life, and it maintains that it is often better to satisfice than to maximize. The content person will make use of practical reason to determine when enough is enough over the course of her or his life, where this is also determined in relation to the other virtues (such as justice and generosity), since the aim is to live the good life, and to not want more than is needed for doing so. The virtue of contentment certainly stands opposed to any tendency toward insatiability, as this means being perpetually discontent. To make the case against insatiability we must make the case for contentment as a virtue, as I have already done in Chapter 1, where I sought to show how it is particularly important for addressing our need to be at home in the world amidst imperfection and human limitation. I will build on this in making my case for the importance of contentment in the economic domain.

One possible virtue that might also be thought to be relevant here is frugality, which is concerned with living simply or minimalistically. In other words, frugality says that "less is more."[16] There is, indeed, a good deal of wisdom in certain conceptions of simple living (especially in an extravagant and hectic age), and this can often go together with the virtue of contentment, though frugality and contentment are not the same. Minimalism is not always conducive to the good life; what we need for the good life is not minimalism, but sufficiency. Some "extravagance" or

---

[16] See Emrys Westacott, *The Wisdom of Frugality: Why Less Is More—More or Less* (Princeton, NJ: Princeton University Press, 2016).

special celebration (e.g., a special feast for a birthday or anniversary or religious holiday) is an important element of a good human life.[17] Of course, it is easy to go too far (and we should remember that fasting and feasting have traditionally gone together). The virtue of contentment is needed to determine what is enough. Therefore, in counteracting the vice of greed in economic life I want to focus on the limiting virtue of contentment.

At this point I want to consider a position that is counter to my own, as well as Aristotle's and the Skidelskys', and to the consensus view of the classical virtue ethics tradition, namely, a defense of the project of becoming as rich as possible, not just for the sake of general human betterment (à la Mandeville, Smith, and Keynes), but also for personal benefit. In his book *Why Not Capitalism?* (which is a rejoinder to Cohen's *Why Not Socialism?*), Jason Brennan writes:

> The more wealth one has, the more one is able to do, and in that sense, the more freedom one has. If we care about people having positive liberty, then we care about making them as wealthy as we can. As the former paramount leader of communist China Deng Xiaoping is thought to have said, "To get rich is glorious!" I agree. In my ideal world, everyone who wants to be is a googolplexinaire.[18]

Brennan thinks that capitalism is preferable to socialism because it makes possible greater prosperity and so greater prospects for positive freedom understood as "the effective power, capacity, or ability to do what one wills" (87). In his more recent book, *Why It's OK to Want to Be Rich*, Brennan says "[money] is freedom," and he thinks people love money because "money can liberate us to have the best chance of leading a life that's authentically our own, of being the authors of our own lives."[19]

---

[17] Recall the discussion of Kass on feasting (from his book *The Hungry Soul*) in Chapter 2.

[18] Jason Brennan, *Why Not Capitalism?* (New York: Routledge, 2014), 87–8.

[19] Jason Brennan, *Why It's OK to Want to Be Rich* (New York: Routledge, 2021), 5–6; see also 29–30, 179–80. In this discussion he says that such money lovers are not necessarily "rapacious and greedy," but still "they want more rather than less, and frequently more than they have" (6). It is not clear how he understands "greedy" here, but the insatiable desire for more money is how I understand greed. The last section of the book before the conclusion is titled "Never Enough." There Brennan says that he thinks "there will always be a point to having *more*" for

In *Why Not Capitalism?* he defines self-authorship as "choosing a conception of the good life and finding the means to achieve that conception."[20] But what does it mean to *choose* a conception of the good life? On what basis are we making such a choice? What Brennan seems to have in mind is having second-order preferences: we have preferences about which of our preferences are central to our life plans. In any case, he is clearly operating with a subjective view of the good life (while also recognizing what Nozick called "side-constraints" on our pursuits in the form of other people's liberty rights).

Brennan is, of course, right that money is helpful for getting many of the things we want. But does it get us happiness? Empirical data on reported happiness suggests that above a certain point (in the United States that point is generally around $75,000 in household income) the correlation is weak, though impoverishment does affect one's happiness negatively.[21] What seems to matter for happiness, then, is that we have *enough*. Moreover, if we understand happiness in terms of *eudaimonia*, that is, an objective understanding of the good life, then we also have reason to reject the project of seeking to be as rich as possible, as we have already seen with Aristotle and the Skidelskys. What matters is that we have enough for living the good life, understood as a normatively higher, nobler, more fulfilling mode of existence that is constituted by virtue, and seeking to be as rich as possible prevents us from living such a life, since the good life is seen here as "a life sufficient unto itself" and the insatiable desire for wealth rejects this. The virtue of contentment helps to know when enough is enough so we do not undermine the good life. It is important to note, however, that the virtue of contentment does not require us to embrace a steady-state economy, since there is often good reason to pursue economic growth, namely, for the benefits it can bring

two reasons: first, because he thinks a steady-state economy is a zero-sum economy where people can only gain at the expense of others, and this leads to antipathy and antagonism between people; and second, because "[the] more one has, the more one can do" (177–9). But, as I explain in the next paragraph, endorsing contentment as a virtue does not entail endorsing a steady-state economy. It is noteworthy that there is not a single entry in the index of Brennan's book for contentment, since the possibility of contentment as a virtue is nowhere discussed.

[20] Brennan, *Why Not Capitalism?*, 92.

[21] See Dan Haybron, *Happiness: A Very Short Introduction* (Oxford: Oxford University Press, 2013), 73–5; Skidelsky and Skidelsky, *How Much Is Enough?*, 102–7.

to ourselves and to others. But it does rule out the unlimited pursuit of economic gain, since this is incompatible with living the good life.

In the next section I will develop a positive vision of the role of economics in the good life, where the focus is on finding a way of living our economic life that contributes to our being properly at home in the world, and the virtue of contentment has a crucial role to play in achieving this. But for now, I want to bring out the importance of contentment through a critique of Brennan's understanding of positive freedom as "the effective power, capacity, or ability to do what one wills," which leads him to affirm the project of seeking to be as rich as possible.

One point to make in response to this view of positive freedom is that it is only as valuable as the ends it enables us to achieve. For Brennan, what it enables us to achieve is whatever it is we happen to want (except, of course, those things that money cannot help us to attain), that is, this view of positive freedom is aimed at the maximal achievement of our preferences and the subjective view of the good life that goes with that. However, as I argued in Chapter 1, if there are no ends of choice of great importance—i.e., objects of intrinsic value that inform an objective conception of the good life and that place constraints on our choices— then this will deflate our sense of the importance of choice. We may still will based on our de facto desires, but this is a disenchanted condition in so far as we accept that as human beings we are meaning-seeking animals who seek to orient our lives toward what is objectively meaningful (or valuable) and worthy of our appreciative attention. As I argued in Chapter 1, an appreciative stance toward what is of intrinsic value in the given world, which places constraints on our desires, provides the necessary background against which genuinely significant choices can be made. If we come to accept the disenchanted view, then we may find, as we saw David Wiggins put it, "a new disquiet" assails our desires: why desire *anything*? We may find ourselves in a debilitating condition, and so lacking positive freedom.

Another point to make here is that there is another conception of positive freedom available (besides Brennan's), which has to do with our effective power to realize an objective conception of the good life in line with a traditional virtue ethics approach. Acquiring the virtues—e.g., justice, generosity, temperance, and contentment—can, in fact, be understood as synonymous with positive freedom as they provide us with the effective power for living out the good life; indeed, they are constitutive of living out the good life. The vices, on the other hand, can be understood as inhibiting this positive freedom, and the vice of greed is a case in point. A lot of traditional warnings about greed speak to how the love of money (for what it helps us to attain: status, security, and the objects of our desire) leads to a kind of enslavement. We can see this, for instance, in the New Testament warnings about attachment to wealth.

Consider the story in the Gospel of Matthew (19:16–26) of the rich young man who approaches Jesus and asks: "Teacher, what good deed must I do to have eternal life?" In his commentary on this story Pope John Paul II notes: "For the young man, the *question* is not so much about rules to be followed, but *about the full meaning of life*," but he correctly senses that "there is a connection between moral good and the fulfilment of his own destiny."[22] Jesus tells the young man that he should keep the commandments (e.g., love your neighbor as yourself; honor your father and mother; do not murder, commit adultery, steal, bear false witness, or covet). The young man says that he has kept these commandments, but he still feels that something is lacking, and he asks Jesus what it is. Apparently sensing his attachment to wealth and knowing that what is most important is that our heart is in the right place (i.e., that our loves are rightly ordered), Jesus replies: "If you wish to be perfect, go, sell your possessions, and give the money to the poor, and you will have treasure in heaven; then come, follow me." Upon hearing this, we are told that the young man "went away grieving, for he had many possessions."[23] Even though Jesus is offering a path to fullness of life, the young man goes away sad, because he is attached to his wealth and presumably to the status and security that it provides, along with the way of life it makes possible. So, given that the young man accepts the vision of fullness of life offered by Jesus, he is not, in fact, free, but enslaved to his wealth.

We also see the concern about enslavement to wealth in the "Sermon on the Mount" in the Gospel of Matthew, when Jesus says: "No one can

---

[22] Pope John Paul II, *Veritatis Splendor* (1993), §§7–8, http://www.vatican.va/content/john-paul-ii/en/encyclicals/documents/hf_jp-ii_enc_06081993_veritatis-splendor.html, accessed July 27, 2021.

[23] The translation I am using is the New Revised Standard Version.

serve two masters...You cannot serve God and wealth" (6:24; see also 19–21). Jesus goes on to warn against overconcern with material security, where a choosing-controlling stance becomes dominant, and we are encouraged instead to take an accepting-appreciating stance by imitating the unconcerned birds of the air and the lilies of the field (6:25–32). This is not to deny a place for human effort and prudence in making a living for ourselves, but we should "strive first for the kingdom of God and his righteousness" (6:33). Contentment is counseled in a number of places in the New Testament. For instance, in 1 Timothy 6, St. Paul says that "there is great gain in godliness [i.e., righteousness] combined with contentment"; however, "those who want to be rich fall into temptation and are trapped by many senseless and harmful desires that plunge people into ruin and destruction. For the love of money is a root of all kinds of evil." Precisely because of how greater wealth enables us to get more of what we want—what Brennan regards as positive freedom—it gives more free rein to our desires, and it can encourage their multiplication in a way that is unconnected with living a good life, and thus it can lead to all kinds of evil and harm and also to our being enslaved to our insatiable desires.[24] True freedom comes rather through contentment.

[24] In Dostoevsky's *The Brothers Karamazov* the character Father Zosima echoes St. Paul in the following teachings that are worth considering here (I had them in mind in my commentary):

> The world has proclaimed the reign of freedom, especially of late, but what do we see in this freedom of theirs? Nothing but slavery and self-destruction! For the world says: "You have desires and so satisfy them, for you have the same rights as the most rich and powerful. Don't be afraid of satisfying them, and even multiplying your desires." That is the modern doctrine of the world. In that they see their freedom. And what follows from this right of multiplication of desires? In the rich, insolation and spiritual suicide; in the poor, envy and murder...They maintain that the world is getting more and more united, more bound together in brotherly communion...Alas, put no faith in such a bond of union. Interpreting freedom as the multiplication and rapid satisfaction of desires, men distort their own nature, for many senseless and foolish desires and habits and ridiculous fancies are fostered in them.... [Instead] of serving the cause of brotherly love and the union of humanity, [they] have fallen on the contrary into dissension and isolation...And therefore the idea of service of humanity, of brotherly love and the solidarity of mankind, is more and more dying out in the world, and indeed this idea is sometimes treated with derision. For how can a man shake off his habits, what can become of him if he is in such bondage to the habit of satisfying the innumerable desires he has created for himself? He is isolated, and what concern has he with the rest of humanity? They have succeeded in accumulating a greater mass of objects, but the joy in the world has grown less.
>
> (*The Brothers Karamazov*, trans. Constance Garnett, rev. Ralph E. Matlaw [New York: Norton, 1976 (1880)], 292–3)

I want to return now to a point in Brennan's remarks that I have not yet commented on, which is when he says: "In my ideal world, everyone who wants to be is a googolplexinaire," that is, *extremely* rich. In fact, given his view of positive freedom, it seems he is committed to saying that everyone should want to be extremely rich. But if everyone became so, then presumably costs would just rise, and this would prevent people from getting whatever they want. The kind of positive freedom that Brennan prizes seems to depend essentially on a significant degree of economic inequality and so significant degrees of difference in positive freedom as he understands it. Hence the desire to be more and more wealthy often goes along with status obsession, which can take on a Promethean aspect of desiring to be godlike in one's effective power to get what one wants, where one ends up seeing the world as one's oyster (to recall Pistol's remark in Shakespeare's *The Merry Wives of Windsor*).[25]

To bring out the connection between the love of wealth and Prometheanism, I want to close this section by reflecting on the Brothers Grimm's fairy tale "The Fisherman and His Wife."[26] In this story a poor fisherman catches a magic flounder, which is actually an enchanted prince, and it asks the fisherman to spare its life. The fisherman does so (since he would not eat a fish that talks), and then he goes home to his wife and tells her about it. She asks why he did not wish for anything and implores him to go back and ask the flounder for a small cottage, so they do not have to live in their squalid hut anymore. Reluctantly, the fisherman goes back to the sea and makes his wife's request to the flounder, which then grants the request. Upon returning home, the fisherman finds his wife in a small cottage, and she seems pleased about it, and the fisherman says: "Now we will live quite contented." But his wife is not content. She keeps sending him back to request something more, and each time the fisherman is more reluctant

---

[25] Brennan in fact endorses Prometheanism. Since he sees increasing wealth as connected to increasing freedom—understood as the effective power to get what one wants—he holds out hope that in the long run "our descendants might live as gods" (*Why It's Ok to Want to Be Rich*, 179).

[26] I am using Margaret Hunt's 1897 translation, reprinted in *The Complete Grimm's Fairy Tales* (New York: Race Point, 2013). The tale was first published by the Brothers Grimm (Jacob and Wilhelm) in 1812.

and protests the impropriety of the request, finding something increasingly sinister about the situation. He is made to request a castle, and then for her to be king, and then emperor, and then Pope. Finally, she is upset that the sun and moon do not rise at her command and so she wishes to be "like unto God," and at this the fisherman is completely horrified. Falling into a rage, she compels him to go to the flounder, and when the fisherman makes the request, the flounder says: "Go to her, and you will find her back again in the dirty hovel." And so the tale ends.

What is the moral of the story? Clearly it teaches us about the insatiability of human desire, especially for wealth, power, and status, and it is meant to show the importance of contentment (the fisherman keeps imploring his wife to contentment). One interesting reading of the ending is that we are "like unto God" (in a proper "godliness" sense) if we are content with what we have, as the fisherman seemed to be (apparently he did not see a compelling reason even for wishing for the cottage to the replace their hut, though arguably here he should not have been content, since a squalid hut can reasonably be regarded as not sufficient).[27] We might then take the message to be something like what is expressed in these remarks from the Talmud: "Who is rich? He who is content with his lot."[28] I would add: provided that it is sufficient for human wellbeing.

Another aspect of the story that it is important to comment on is the way it reveals—as many of the Brothers Grimm's fairy tales do—the darker, more troublesome aspects of the human psyche. We see this in the way in which the fisherman finds something increasingly sinister in his wife's requests for wealth, power, and status. Cora Diamond describes how the story conveys a kind of "unapproachable evil" in the tendency of human willfulness that we see in the wife, where "she begins by wishing her way into a decent little [cottage], and ends by wishing to be God, finding it an offense that the sun and moon should rise whether she wills

---

[27] This reading is strongly suggested in Patricia Polacco's version of the fairy tale, *Luba and the Wren* (New York: Philomel, 1999).

[28] *Mishnah, Avot* 4:1; translation in Ben Zion Bokser, *The Wisdom of the Talmud: A Thousand Years of Jewish Thought* (New York: Citadel Press, 1951), 125. Thanks to Daniel Goodman for help with this reference.

it or not."[29] There is something horrifying (an "unapproachable evil") in the prospect of a kind of human willfulness that recognizes no constraints whatsoever on what one wills and so is willing to violate anything of intrinsic value that stands in the way. In fact, in the story we see an increasing disregard of the fisherman by his wife as she increases in her status. This suggests a potential danger in the desire for status: even where it does not lead to the extremes of the fisherman's wife, it can still lead to forms of disregard for other human beings, as one sees oneself above others. The Promethean stance of the fisherman's wife in "playing God," it seems, is an inherent tendency in the insatiable desire for wealth, power, and status; it is the most extreme outcome of this desire if it is not curbed by not only the limiting virtue of contentment but also the limiting virtue of humility. As discussed in Chapter 1, the virtue of humility ensures that we recognize and live out our proper place in the scheme of things, where we do not think too much of ourselves and recognize proper limits on our will.

It should be noted that there is something strange about the fisherman's wife thinking so highly of herself, since her status was not a result of her own efforts, but rather it was the result of her being granted wishes by a magical fish; we might say that this was merely a matter of good luck. However, even where effort is involved, there can still be a problematic form of hubris, which Michael Sandel calls "meritocratic hubris." By this he means the tendencies of economic and social-status "winners" to take too much pride in their success and to forget the good fortune that helped them get to where they are. They smugly believe that they fully deserve their status, and also that the "losers" at the bottom of the economic and social-status ladder are deserving of theirs. To counteract such hubris and cultivate proper humility and social solidarity, Sandel maintains that we need "a lively sense of the contingency of our lot," where we think, "There, but for the grace of God, or the accident of fortune, go I."[30] I think this is right, but I would add that a lively,

---

[29] Cora Diamond, "Ethics, Imagination, and the Method of Wittgenstein's *Tractatus*," in *The New Wittgenstein*, ed. Alice Crary and Rupert Read (New York: Routledge, 2000), 166.

[30] Sandel, *The Tyranny of Merit*, 25. Recall his similar remark (cited in Chapter 1) in *The Case against Perfection*: "An appreciation of the giftedness of life constrains the Promethean project and conduces to a certain humility" (27).

reverential sense of intrinsic human dignity is also important. As discussed in Chapter 1, humility and reverence are closely linked virtues, as humility makes us receptive to values not of our own making (including the reverence-worthy) and reverence—in recognizing limits on our will—helps to define our proper place in the scheme of things.

Home Economics

The idea that we should pursue more and more wealth fits with the idea of *homo economicus*, which is the view of human beings underlying the standard "rational choice" model of human behavior among contemporary economists.[31] The idea here is that human beings are preference (or utility) maximizers who are looking for the most efficient means for maximally realizing their preferences, where determining efficiency requires a cost–benefit analysis. The view does not say that human beings cannot act for ethical reasons, but these must be regarded as preferences among other preferences. This distorts the nature of ethical reasons which recognize normative standards (e.g., the virtues) for what we ought to desire, and so should not be regarded as mere preferences among other preferences. The virtue of contentment also has no place in the view of a human being as *homo economicus*, given the maximizing assumption of this view.

There is a lot about human motivation that *homo economicus* does not capture. One important human motivation that it neglects is the need for home (i.e., belonging), which connects with the virtue of contentment, since, as I have construed it, it is concerned with finding a way of being *at home* in the world. Roger Scruton writes: "Our condition is not that of *Homo oeconomicus*, searching in everything to satisfy his private desires. We are home-building creatures, cooperating in the search for intrinsic values, and what matters to us are the ends, not the means of our

---

[31] See Gary S. Becker, *The Economic Approach to Human Behavior* (Chicago: University of Chicago Press, 1976); for a good description of *homo economicus*, see Mary L. Hirschfield, *Aquinas and the Market: Toward a Humane Economy* (Cambridge, MA: Harvard University Press, 2018), 39–42.

existence."[32] Our being home-building creatures is also connected to what Scruton calls *oikophilia*, which arises when we are at home in the world:

> Human beings, in their settled condition, are animated by *oikophilia*: the love of the *oikos*, which means not only the home but the people contained in it, and the surrounding settlements...The *oikos* is the place that is not just mine and yours but *ours*.... Virtues like thrift and self-sacrifice, the habit of offering and receiving respect, the sense of responsibility—all those aspects of the human condition that shape us as stewards and guardians of our common inheritance—arise through our growth as persons, by creating islands of value in the sea of price. To acquire these virtues we must circumscribe the "instrumental reasoning" that governs the life of *Homo oeconomicus*. We must vest our love and desire in things [of] intrinsic, rather than ... instrumental, value, so that the pursuit of means can come to rest, for us, in a place of ends. That is what we mean by settlement: putting the *oikos* back in the *oikonomia*.[33]

As alluded to here, "economics" comes from the Greek word *oikonomia*, which is standardly translated as "household management." The best-known account of *oikonomia* is found in Book I of Aristotle's *Politics*. Mark Shiffman provides the following helpful summary:

> When he attempts to clarify the proper conduct of *oikonomia* in *Politics* I.8–11, Aristotle is at pains to distinguish the art of household management from the art of acquisition. The former looks to obtaining, preserving, and using "those goods a store of which is both necessary for life and useful for partnership in a city or a household." The acquisitive art, on the other hand, seeks to discern "what and how to exchange in order to make the greatest profit." Accordingly,

---

[32] Scruton, *How to Be a Conservative*, 119. See also Christopher Toner, "Home and Our Need for It," *Journal of Philosophical Research* 44 (2019): 251–72.

[33] Scruton, *How to Be a Conservative*, 24–5. The concept of *oikophilia* is explored at length in Scruton's *How to Think Seriously about the Planet: The Case for an Environmental Conservatism* (Oxford: Oxford University Press, 2012).

oikonomia recognizes limits to acquisition, since it takes its measure by the standard of self-sufficiency with a view to a good life, which means a life embodying virtue and friendship. Those intent on profit, on the other hand,... since "they are serious about living, but not about living well; and since that desire of theirs is without limit, they also desire what is productive of unlimited things."[34]

What I want to suggest is that in order to realize the proper place of economics in the good life, we need to recover a "home economics" (which would put the *oikos* back in *oikonomia*), that is, we need a way of living our economic life that contributes to our being properly at home in the world, rather than causing alienation, as it too often does in our age.[35] Scruton refers to human beings in their settled condition, but one of the challenges of our capitalist economic order is that it encourages unsettling tendencies. This was identified in Karl Marx's famous lament about the effects of capitalism in *The Communist Manifesto*: "All fixed, fast-frozen relations, with their train of ancient and venerable prejudices and opinions, are swept away, all new-formed ones become antiquated before they can ossify. All that is solid melts into air, all that is holy is profaned."[36] Marx's lament here is essentially a conservative lament, and his critique of the destructive and unsettling tendencies of capitalism has been shared by a number of traditional conservatives.[37] But Marx, of course, went on to give radical proposals, such as the abolition of private property, for achieving his communist utopia.[38] Marx is right in his lament, but wrong in his proposed solution. For one thing, we should

[34] Mark Shiffman, "The Rediscovery of *Oikonomia*," in *The Humane Vision of Wendell Berry*, iBook edn, ed. Mark T. Mitchell and Nathan Schlueter (Wilmington, DE: ISI Books, 2011), 354–5.

[35] Wendell Berry has a collection of essays titled *Home Economics* (New York: North Point Press, 1987), and he says of the title: "Its redundancy seems to acknowledge that what passes for economics...has strayed far from any idea of home" (x). I will say more about Berry's work shortly.

[36] Karl Marx and Friedrich Engels, *The Communist Manifesto* (1848), in *Karl Marx: Selected Writings*, ed. David McLellan, 2nd edn (Oxford: Oxford University Press, 2000), 248.

[37] See Peter Kolozi, *Conservatives against Capitalism: From the Industrial Revolution to Globalization* (New York: Columbia University Press, 2017); Patrick J. Deneen, *Why Liberalism Failed* (New Haven, CT: Yale University Press, 2018); and Eugene D. Genovese, *The Southern Tradition: The Achievements and Limitations of an American Conservatism* (Cambridge, MA: Harvard University Press, 1994).

[38] See Marx and Engels, *The Communist Manifesto*, 256–62.

distrust such utopian schemes, especially regarding their feasibility (as already discussed in Chapter 3). But Marx's solution is also not desirable: we should rather affirm the importance of private property for settlement or being at home in world, that is, it is important to have property of our own (a house, land, etc.), marked off from public spaces and public demand, which enables us to belong to a particular place and encourages a sense of responsibility.[39] The proper response to the unsettling tendencies of capitalism is *economic decentralization* (accompanied by political decentralization of the sort that I defended in Chapter 3), which means endorsing policies that help to disperse property widely and sufficiently, rather than having it concentrated in the hands of a few, and promoting and protecting local economies.[40] This is needed for a truly free market, since a market is not truly free when large corporations control it.

In order to advance my view of home economics (or *oikonomia*) I want to discuss the work of Wendell Berry. In his book *The Unsettling of America*, Berry describes how the dominant tendency in American history has been toward *displacement*:

[39] This defense of private property is suggested by G. W. F. Hegel when he says that "property is the *embodiment* of personality," and when he also says (rather opaquely): "A person must translate his freedom into an external sphere in order to exist as Idea" (*The Philosophy of Right*, trans. T. M. Knox [Oxford: Oxford University Press, 1967 (1821)], §§51, 41). I will return shortly to discuss how Hegel thinks that through work we can make a home in the world. My view here is also influenced by Aristotle's defense of private property (against the proto-Marxist suggestion of Plato for its abolition). Aristotle says: "when care for property is divided up, it leads not to those mutual accusations [that would occur in communal ownership schemes], but rather to greater care being given, as each will be attending to what is his own" (P II.5, 1263a26–8). This personal responsibility for property ultimately serves the common good by allowing for greater productivity. Aristotle goes on to say: "Besides, to regard a thing as one's own makes an enormous difference to one's pleasure. For the love each person feels for himself is no accident, but is something natural" (II.5, 1263a40–1263b1). He says that selfishness is rightly criticized, but he thinks that private property makes possible the virtue of generosity (II.5, 1263b7–14). Aristotle also thinks that it is "fair to mention" that communal ownership schemes are "totally impossible" (II.5, 1263b27–8).

[40] This position is sometimes called "distributism," though the label is not very clear. I think it is better to speak of economic decentralization. This position has been advocated by modern Catholic social teaching, starting with Pope Leo XIII's 1891 encyclical *Rerum Novarum*, which influenced the economic writings of Hilaire Belloc and G. K. Chesterton. Other prominent advocates for economic decentralization in this tradition are Wilhelm Röpke (*A Humane Economy: The Social Framework of the Free Market* [Chicago: Regnery, 1960]), E. F. Schumaker (*Small Is Beautiful: Economics as if People Mattered* [New York: Harper, 1973]), and Wendell Berry (whose work I will be discussing, and who sees himself in the tradition of Thomas Jefferson on economic decentralization). The Skidelskys also endorse a version of this view; see Skidelsky and Skidelsky, *How Much Is Enough?*, 161–2 and ch. 7.

As a people, wherever we have been, we have never really intended to be.... The earliest explorers were looking for gold, which was... always somewhere farther on.... Once the unknown of geography was mapped, the industrial marketplace became the new frontier, and we continued, with largely the same motives and with increasing haste and anxiety, to displace ourselves.[41]

However, there has also been another tendency that so far has been "the weaker tendency, less glamorous, certainly less successful," namely, the "tendency of settlement, of domestic permanence," that is, "the tendency to stay put, to say, 'No farther. This is the place'" (4, 13). Berry sees here two basic stances toward place: the exploiter stance and the nurturer stance. These stances can be seen as connected to the two fundamental existential stances that I discussed in Chapter 1, namely, the choosing-controlling stance and the accepting-appreciating stance toward the given. The exploiter stance is a certain extreme direction of the choosing-controlling stance where we see the world as our oyster to do with as we will; in other words, it is a Promethean stance of mastery. The nurturer stance, on the other hand, can be seen as following from a proper accepting-appreciating stance toward the given world: when we properly appreciate something (or someone), we want to protect, preserve, and nurture it (or him or her). Berry writes:

I conceive a strip-miner to be a model exploiter, and as a model nurturer I take the old-fashioned idea or ideal of a farmer.... The standard of the exploiter is efficiency; the standard of the nurturer is care. The exploiter's goal is money, profit; the nurturer's goal is health—his land's health, his own, his family's, his community's, his country's.... The exploiter wishes to earn as much as possible by as little work as possible; the nurturer expects, certainly to have a decent living from his work, but his characteristic wish is to work *as well* as possible. (7–8)

[41] Wendell Berry, *The Unsettling of America: Culture & Agriculture* (New York: Avon Books, 1977), 3.

The difference here over work is of central importance. We have seen Keynes suggest that work is an affliction that is ideally overcome, and he is hardly alone in this. Indeed, the biblical story of the Garden of Eden suggests that work is a curse for disobedience. However, in his reading of the story of the Garden of Eden, G. W. F. Hegel maintains that work is not only a curse but also a way of redemption. The disobedience of Adam and Eve in eating of the "tree of the knowledge of good and evil" and their expulsion from the garden symbolizes our emergence into rational, ethical, self-conscious agency, and Hegel says that "if [work] is the result of the disunion, it is also the victory over it": "the second harmony must spring from the labour and culture of the spirit."[42] In other words, it is through work that we can overcome alienation and make a home in the world and belong. As Hegel puts it elsewhere: "[Man] brings himself before himself by practical activity, since he has the impulse, in whatever is directly given to him, in what is present to him externally, to produce himself and therein equally to recognize himself."[43] In other words, through our work we can see ourselves in the world, and thus overcome the strict dualism between subject and object. Later Hegel writes about how through work a human being "humanizes his environment": "Only by means of this effectual activity is he no longer merely in general, but also in particular and in detail, actually aware of himself and at home in his environment" (256). Hegel gives emphasis here to our acting upon the world and making ourselves at home within it, but it is also important to recognize the role of the accepting-appreciating stance toward what is of intrinsic value in the given world (or what in religious terms we might call receptivity to grace, that is, given unmerited good) for becoming at home in the world. I will come back to discuss this matter more in the final section.

It is helpful here also to consider the modern tradition of Catholic social teaching, which has reflected extensively on the dignity of work and its proper place in human life, starting with Pope Leo XIII's *Rerum Novarum* (1891; English title: "The Condition of Labor"). Consider in

---

[42] G. W. F. Hegel, *Hegel's Logic: Being Part One of the* Encyclopedia of the Philosophical Science, trans. William Wallace (Oxford: Oxford University Press, 1975 [1830]), §24. See n. 18 in Chapter 2.

[43] G. W. F. Hegel, *Aesthetics: Lectures on the Fine Arts*, trans. T. M. Knox (Oxford: Oxford University Press, 1975 [1835]), 31.

particular Pope John Paul II's encyclical *Laborem Exercens* (1981; English title: "On Human Work"), where he makes three important points about the dignity of human work (§§9–10).[44] First, through work we realize the biblical mandate that human beings should have "dominion" over the earth (Genesis 1:26) in the manner that is proper to us and compatible with reverence for God and the created order. This is part of our dignity in being made in the image and likeness of God, since through our work we imitate God's creative work in creation. Of course, from Jewish and Christian perspectives, we also need to imitate God in practicing the Sabbath, which for human beings helps us to recognize limits on human dominion. I will return in the final section to discuss the importance of a Sabbath-orientation for everyone, regardless of religious beliefs. But it is important to see here that there is a proper kind of dominion, where we adopt a choosing-controlling stance within proper limits. Like Hegel, John Paul II does not think that the toil of work should be regarded merely as a curse: "Work is a good thing for man—a good thing for his humanity—because through work man *not only transforms nature*, adapting it to his own needs, but also *achieves fulfillment* as a human being and indeed in a sense becomes 'more a human being.'"[45]

[44] See Pope John Paul II, *Laborem Exercens* (1981), reprinted in *Catholic Social Thought: The Documentary Heritage*, ed. David J. O'Brien and Thomas A. Shannon (Maryknoll, NY: Orbis Books, 1992).

[45] David Wiggins advances a similar view when he draws on Aristotle to go beyond him in discussing the importance of ordinary work:

> As it figures in *Nicomachean Ethics* Book 1, Chapter 7, isn't *ergon* what … we call *work* … ? A human being pursues the human good that is happiness, Aristotle will now appear to suggest, by doing the work … that is proper to a human person. Loosening the tie to Aristotle, one might say we pursue happiness by doing well the work that is proper to the particular person we are, given our abilities, our reasonable predilections, our situation and our commitments …. It is in doing that sort of work that a man or woman has a life worth living …. Acts or activities that apply what [Aristotle] calls a rational principle aim at something worthwhile by drawing upon faculties and dispositions whose exercise gives pleasure … to the doer and enlarges also—here I reach beyond Aristotle—the doer's understanding of the realities we inhabit. That is to say that the exercise of these faculties or dispositions affords both practical understanding of those realities and the satisfactions that we attain by learning to wrestle or struggle with them. On these terms … engagement with the human *ergon*, taken as I am suggesting we take this, must be as important to a person as anything can be to him or her …. [It] will shape our understanding of life, leisure and repose, and mould our aspirations for the calling or occupation by which we shall subsist and get the necessities of life itself.
>
> ("Work, Its Moral Meaning or Import," *Philosophy* 89:3 [2014]: 478–80)

In other words, work is part of our dignity and fulfillment because through it we exercise our rational agency in meeting our needs, including, I would add, our need for belonging and being at home in the world. John Paul II also notes that there are ways that one can be degraded through work in bad conditions, which means that we have an obligation to protect against these bad conditions to ensure respect for the dignity of human work.

The second point about the dignity of work that John Paul II makes is that work constitutes the "foundation of family life," both in that through work one seeks to provide for one's family and in that the family is the "first school of work." The third and final point that John Paul II makes about the dignity of work is that through it we contribute to the common good of our community and also of humanity as a whole. These last two points about the dignity of work identify two important ways that work contributes to human belonging, and all three points can be seen as connected to a conception of work as a calling or vocation through which we pursue the good life both for ourselves and for those with whom we share in community.

Berry holds similar views on work to those we have seen expressed by Hegel and Catholic social teaching. He notes that many have regarded work as beneath human dignity, as something to be escaped from as far as possible, and so they have failed to appreciate the dignity of work and its importance in human fulfillment. However, he questions whether we can escape from work with impunity, and he thinks that ancient wisdom teaches that we cannot:

> It tells us that work is necessary to us, as much a part of our condition as mortality; that good work is our salvation and our joy; that shoddy...work is our curse and our doom. We have tried to escape the sweat and sorrow promised in Genesis—only to find that, in order to do so, we must forswear love and excellence, health and joy.[46]

We saw that Berry says that the nurturer's goal is health, which elsewhere he notes "comes from the same Indo-European roots as 'heal', 'whole',

---

[46] Berry, *The Unsettling of America*, 12.

and 'holy'": *"To be healthy is literally to be whole; to heal is to make whole."*[47] We need to be made whole because, as we emerge from childhood innocence into the ethical self-consciousness of adulthood, we are painfully instructed by experience about the realities of division within ourselves, with others, and with the world. But, Berry says:

> if our culture works in us as it should, then we do not age merely into disintegration and division, but that very experience begins our education, leading us into knowledge of wholeness and of holiness.... [We] are led out of our lonely suffering and are made whole.[48]

A key work of culture here is to instruct us about the importance of recognizing limits in human life for avoiding dehumanization and for realizing our fulfillment as human beings:

> A culture is not a collection of relics or ornaments, but a practical necessity, and its corruption invokes calamity. A healthy culture is a communal order of memory, insight, value, work, conviviality, reverence, aspiration. It reveals the human necessities and the human limits. It clarifies our inescapable bonds to the earth and to each other. It assures that the necessary restraints are observed, that the necessary work is done, and that it is done well.[49]

The realities of our economic life, however, pose significant challenges for the "labour and culture of the spirit" (as Hegel puts it) in the task of helping us to become whole and at home in the world. For many people today work is, in fact, a source of alienation rather than a way of overcoming it because of how they feel disconnected from its product and from the communities that work should serve—a condition that Marx famously described in his account of alienated labor.[50] We need,

---

[47] Wendell Berry, "Health Is Membership," in *The Art of the Commonplace: The Agrarian Essays of Wendell Berry* (Washington DC: Shoemaker & Hoard, 2002), 144.

[48] Berry, "Health Is Membership," 145.   [49] Berry, *The Unsettling of America*, 43.

[50] See Karl Marx, "Economic and Philosophical Manuscripts" (1844), in *Karl Marx: Selected Writings*.

therefore, to find a way to live our economic life that can help us to feel at home in the world, and this requires recognizing economic limits.

In his essay "Faustian Economics: Hell Hath No Limits," Berry notes that once greed has been made an honorable motive, as it has become in our prodigal age, then you have an "economy without limits":

> In keeping with our unrestrained consumptiveness, the commonly accepted basis of our economy is the supposed possibility of limitless growth, limitless wants, limitless wealth, limitless natural resources, limitless energy, and limitless debt. The idea of a limitless economy implies and requires a general doctrine of human limitlessness: *all* are entitled to pursue without limit whatever they conceived as desirable.[51]

However, this does not square with our actual human limits and with the limits of natural resources. Berry writes:

> We are...coming under pressure to understand ourselves as limited creatures in a limited world. This constraint, however, is not the condemnation it may seem. On the contrary, it returns us to our real condition and to our human heritage, from which our self-definition as limitless animals has for too long cut us off. Every cultural and religious tradition that I know about, while fully acknowledging our animal nature, defines us specifically as humans—that is, as animals...capable of living not only within natural limits but also within cultural limits, self-imposed. As earthly creatures, we live, because we must, within natural limits...But as humans, we may elect to respond to this necessary placement by the self-restraints implied in neighborliness, stewardship, thrift, temperance, generosity, care, kindness, friendship, loyalty, and love.... [Our] cultural tradition is in large part the record of our continuing effort to understand ourselves as being specifically human: to say that, as humans, we must do certain things and we must not do certain things. We must have limits or we will cease to exist as humans; perhaps we will cease to exist, period.... [Our] human and

---

[51] Wendell Berry, "Faustian Economics: Hell Hath No Limits," *Harper's Magazine* (May 2008): 36.

earthly limits, properly understood, are not confinements but rather inducements to formal elaboration and elegance, to *fullness* of relationship and meaning.   (37–9, 41)

Berry's remark here that "[we] must have limits or we will cease to exist as humans" (i.e., humans in the normative sense of being fully human) expresses well the overall message of this book. In another essay titled "The Thought of Limits in a Prodigal Age," Berry similarly writes:

Enduring structures of household and family life, or the life of a community or the life of a country, cannot be formed except within limits. We must not outdistance local knowledge and affection, or the capacities of local persons to pay attention to details, to the "minute particulars" *only* by which, William Blake thought, we can do good to one another. Within limits, we can think of rightness of scale. When the scale is right, we can imagine completeness of form.[52]

As suggested here, one of the key self-imposed limits that we need is loyalty to our particular place and the particular people with whom we live in community in this place, from family and friends to neighbors and fellow citizens. In his Jefferson Lecture, "It All Turns on Affection," Berry makes a distinction between "boomers" and "stickers" that is related to his distinction between exploiters and nurturers: "boomers" are "those who pillage and run," who want "to make a killing and end up on Easy Street," and they are "motivated by greed, the desire for money, property, and therefore power"; by contrast, "stickers" are "those who settle, and love the life they have made and the place they have made it in," and they are "motivated by affection, by such love for a place and its life that they want to preserve and remain in it."[53]

---

[52] Wendell Berry, "The Thought of Limits in a Prodigal Age," in *The Art of Loading Brush: New Agrarian Writings* (Berkeley, CA: Counterpoint, 2017), 55.

[53] Wendell Berry, *It All Turns on Affection: The Jefferson Lecture & Other Essays* (Berkeley, CA: Counterpoint, 2012), 10–11. Berry draws this distinction from Wallace Stegner, *Where the Bluebird Sings to the Lemonade Springs: Living and Writing in the West* (New York: Random House, 1992).

As already suggested, we seem to live in an age defined by the restless pursuit of material improvement, power, and status. However, at the same time we have a fundamental human need for settlement, that is, for being at home in the world. To be a "sticker" in our age clearly in some ways will involve going against the grain and will be difficult. And so it seems that we might need something like a "vow of stability," such as Benedictine monks and nuns take.[54] A vow of stability can be understood precisely as a commitment of loyalty to a particular place and the particular people with whom we share in community in this place. I do not think it should be understood as having quite the strength of a traditional marriage vow in terms of indefeasibility, since one may have to move for reasons of economic necessity or doing what is best, all things considered, for one's family. But the idea here is about our orientation: if we are to live well as the home-seeking, home-building, and home-loving creatures that we are, then we need to find somewhere to belong, we need to recognize ties that bind, and we should not always be looking for the next best economic or status-enhancing opportunity.

It should be acknowledged that the overall tendencies toward economic centralization today—as seen in the power of large corporations, often operating on an international scale, and in the growth of big cities as economic centers and the economic hollowing out of small towns— pose significant challenges to the vision that I have put forward of home

---

[54] See St. Benedict, *The Rule of St. Benedict in English*, ed. Timothy Fry, OSB (Collegeville, MN: Liturgical Press, 1981 [*c*.535–40]), ch. 4. Thomas Merton writes:

> [Saint Benedict] introduced this vow into his Rule precisely because he knew that the limitations of the monk, and the limitations of the community he lived in, formed a part of God's plan for the sanctification both of individuals and communities. By making a vow of stability the monk renounces the vain hope of wandering off to find a "perfect monastery." (*The Sign of Jonas* [New York: Harvest Book, 1953], 9–10)

Sr. Jane Michele McClure, OSB writes:

> Stability means that the monastic pledges lifelong commitment to a particular community. To limit oneself voluntarily to one place with one group of people for the rest of one's life makes a powerful statement. Contentment and fulfilment do not exist in constant change; true happiness cannot…be found anywhere other than in this place and this time. For Benedictines, the vow of stability proclaims rootedness, "at-homeness."
>
> ("Benedictine Way of Life,"; http://stellamaris.nsw.edu.au/wp-content/uploads/ 2015/06/benedictine-way-of-life.pdf, accessed July 28, 2021)

economics, given its commitment to economic decentralization. But the vision is not utopian, because it has been realized throughout human history, including in the modern world, though increasingly less so, it seems. In line with a longstanding tradition of economic thought going back to Aristotle, I have put forward the vision of home economics, as a vision of the proper role of economics in the good life, as something that we should strive to achieve as far as possible so that we can be at home in the world as far as possible. And given that we are limited creatures living in a limited world, we may ultimately have no other choice. As Berry puts it:

> It has seemed to me less a choice than a necessity to oppose the boomer enterprise with its false standards and its incomplete accounting, and to espouse the cause of stable, restorative, locally adapted economies of mostly family-sized farms, ranches, shops, and trades. Naïve as it may sound now, within the context of our present faith in science, finance, and technology, this cause nevertheless has an authentic source in the sticker's hope to abide in and to live from some chosen and cherished small place—which is the agrarian vision that Thomas Jefferson spoke for, a sometimes honored human theme, minor and even fugitive, but continuous from ancient times until now. Allegiance to it, however, is not a conclusion but the beginning of thought.[55]

Berry's point here about how allegiance to home economics (*oikonomia*) is not a conclusion but the beginning of thought is an important one. It will require practical wisdom to discern how best to realize it as far as possible within the circumstances of each of our lives.

A Sabbath-Orientation

In Chapter 1 I discussed several reasons why the accepting-appreciating stance toward the given world needs to be regarded as primary over the

---

[55] Berry, *It All Turns on Affection*, 18–19.

choosing-controlling stance. One reason for this is because we first need to be properly responsive to what is of value in the given world in order to know how to act (or not act). In other words, the accepting-appreciating stance should inform when and how we take up the choosing-controlling stance. Another point that I made is that in virtue of the limits of our existence we need to recognize that a state of perfection will never be realized through our efforts, and so we need a way of living with and being at home amidst imperfection, which means that we need a way of coming to see life in the world as good and worth affirming despite the ill. Finally, I mentioned that in an important sense our achievements are not really complete without our appreciation of them, and I illustrated this by appealing to the creation story in Genesis, where God creates the world in six days and then *completes* his creation through appreciating it—where he contemplatively beholds it as "very good"—and resting on the seventh day. The practice of the Sabbath imitates God in creation: the Sabbath completes our own work through restful appreciation of this work as well as the world in which we live. In this final section I want to discuss the practice of the Sabbath in more detail and to suggest its relevance for everyone, regardless of religious beliefs. Indeed, I want to show how it is important not only for completing our achievements but also for cultivating proper responsiveness to what is of value in the given world and for finding a way of coming to see life in the world as good and worth affirming despite the ill. Reflecting on the Sabbath is not only a fitting way to conclude this chapter on economic limits, but it is also a fitting way to conclude the book because it returns us to the discussion with which this book began: the two fundamental existential stances—i.e., the choosing-controlling stance and the accepting-appreciating stance toward the given world—and how we should understand the relationship between them.

When we think of the Sabbath, we think of rest, or a break from normal activities (as in a sabbatical). For Jews and Christians, the Sabbath (or *Shabbat*) is a specific weekly *day* of rest (Saturday for Jews, though beginning Friday evening; Sunday for Christians), as one of the Ten Commandments instructs: "Observe the sabbath day and keep it holy" (Deuteronomy 5:12; Exodus 20:8 says: "Remember the sabbath day, and keep it holy"). However, Samuel Fleischacker notes in his reflections on the Jewish Sabbath:

> Shabbat is defined negatively, in the first instance. "Shabbat" comes from a word that literally means "ceasing" or "refraining" rather than "resting"; it is also known, among both Jews and non-Jews, primarily for what one may not do on it rather than what one may do; and the experience of shabbat, while by no means confined to these prohibitions, is pervasively shaped by them.[56]

The main prohibitions of the traditional Jewish Sabbath are that one is not to cook, write, use money, manipulate electrical appliances, or use a vehicle. As Fleischacker puts it, "shabbat is meant to relieve us from concern for our material needs," though even beyond that, in practice, these prohibitions "keep sabbath-observant Jews from their professional work and from everyday leisure activities like watching TV or going to the movies" (118). But we miss something crucial about the Sabbath if we only focus on the restrictions, for the restrictions make room for other important activities, namely, spending time with family and friends, sharing meals, singing, reading, praying, attending religious service, worshipping, and engaging in contemplation. Fleischacker notes that the Jewish tradition has, in fact, developed a positive program of these Sabbath activities.

We can see here what the Sabbath prescribes is in fact a form of leisure, where leisure is understood as a break from work that is concerned with meeting our material needs and/or seeking status and involves engaging in intrinsically valuable activities for their own sake, including especially contemplation, that is, appreciative attention to what is of intrinsic value. I want to suggest that there is a way that anyone, regardless of religious belief, can practice the Sabbath, as a day of leisured rest—or it could be longer or shorter than a day—that resembles the traditional Jewish program for the Sabbath. Focusing on the Sabbath

---

[56] Samuel Fleischacker, "The Jewish Sabbath as a Spiritual Practice," in *Spirituality and the Good Life: Philosophical Approaches*, ed. David McPherson (Cambridge: Cambridge University Press, 2017), 118.

is helpful here because, as we have seen, some leisure advocates—Aristotle, Keynes, and to some extent the Skidelskys—have denigrated or downplayed the importance of work. However, the Sabbath preserves an integral connection with work in being its *completion* or *fulfillment*, and in doing so it recognizes the dignity and meaningfulness of human work. You cannot have the Sabbath without work, though unfortunately people do have work without the Sabbath.

When we practice the Sabbath in the right spirit, we develop what we can call a "Sabbath-orientation" for our lives, where the fulfillment of all our work resides in the restful and appreciative enjoyment of the good things of life, including our work.[57] Here one ceases for a time from the choosing-controlling stance and instead adopts an accepting-appreciating stance toward what is of value in the given world, and also, if one is a theist, toward its ultimate creative source (God). We can see then that the Sabbath rightly practiced not only completes or fulfills our work through appreciation of this work, but it also helps us to orient our lives better outside of the Sabbath, since during the Sabbath we engage in contemplation—i.e., appreciative attention to what is of value—and reflect on what is most important. Iris Murdoch—though not discussing the Sabbath specifically—discusses the significance of appreciative attention for the ethical life, saying: "[We] can all receive moral help by focusing our attention upon things which are valuable;... [Our] ability to act well 'when the time comes' depends partly, perhaps largely, upon the quality of our habitual objects of attention."[58] In other words, acting well depends on knowing what values are at stake in our lives and which are most important, and whether we know this depends on our habits of appreciative attention. Murdoch cites St. Paul:

[57] Norman Wirzba uses the phrase "sabbath-orientation" in *Food and Faith: A Theology of Eating* (Cambridge: Cambridge University Press, 2011), 43–8. See also Wirzba, *Living the Sabbath: Discovering the Rhythms of Rest and Delight* (Grand Rapids, MI: Brazos Press, 2006). Fleischacker similarly writes of a "shabbat-consciousness" that pervades one's whole life ("The Jewish Sabbath as a Spiritual Practice," 121), and he also discusses how the Sabbath completes our activities (122–3, 127). He describes what he calls the "creative rest" of the Sabbath as follows: "Shabbat... becomes a condition for a thing to have an intrinsic value, an inherent beauty" (127). I think this puts the matter too much in a constructivist framework; we should say instead that the Sabbath is a condition for *properly recognizing* the intrinsic value of things.

[58] Iris Murdoch, "On 'God' and 'Good,'" in *Existentialists and Mystics: Writings on Philosophy and Literature*, ed. Peter Conradi (New York: Penguin, 1997), 345.

"whatsoever things are true, whatsoever things are honest, whatsoever things are just, whatsoever things are pure, whatsoever things are lovely, whatsoever things are of good report; if there be any virtue, and if there be any praise, think on these things" (Philippians 4:8; KJV).[59]

The idea of a "Sabbath-orientation" can thus be understood to have a double meaning: in our work we should be oriented toward the Sabbath as its proper completion or fulfillment, but also through the contemplation we engage in when rightly practicing the Sabbath we can better orient our lives toward the good. One important aspect of this orientation toward the good that the Sabbath fosters is the appreciation of and respect for limits. As Rabbi David Hartman writes: "the Sabbath develops the characteristic of gratitude, the sense that life is a gift, and the need to give up the longing for absolute power."[60] Fleischacker also notes how the practice of the Sabbath helps to curb idolatrous tendencies to think too highly of ourselves, to worship ourselves and our own projects, and at the extreme, we can add, to "play God" in thinking that we are masters of the given world.[61] The practice of the Sabbath can thus be understood as a practice of humility where we seek to recognize and live out our proper place in the scheme of things by ceasing for a time from the choosing-controlling stance, and instead adopting an accepting-appreciating stance. In an essay commenting on Hartman's work, Michael Sandel remarks: "the celebration of human finitude [inherent in keeping the

---

[59] See my discussion of the contemplative life in ch. 5 of *Virtue and Meaning*, where there is some overlap with the discussion here, but the place of contemplation in human life is discussed in more detail.

[60] David Hartman, *A Heart of Many Rooms: Celebrating the Many Voices within Judaism* (Woodstock, VT: Jewish Lights Publishing, 1999), 201–2. Earlier Hartman writes:

> [On] the Sabbath, a person may not stand over and against the universe as a Promethean figure.... The halakhic notion of the holiness of the Sabbath aims at controlling the impulse to mastery by setting limits to human dominance of nature. Nature is transformed from an "it" into a "thou"; the world is no longer the object of human gratification. The halakhic norms of the Sabbath affirm the value of existence outside of an anthropocentric perspective.... I stand silently before nature as before a fellow creature and not as before a potential object of my control. By forcing us to experience the meaning of being creatures of God, the Sabbath aims at healing the human grandiosity of technological arrogance.   (77–8)

Elsewhere Hartman also writes: "On the Sabbath, Jews celebrate God as the Creator.... Awe, wonder, and humility are expressed by giving up mastery and control over the world for a day. Nature is not our absolute possession" (*A Living Covenant: The Innovative Spirit in Traditional Judaism* [New York: Free Press, 1985], 260).

[61] See Fleischacker, "The Jewish Sabbath as a Spiritual Practice," 122, 125, 128, 131.

Sabbath] means that [one] can affirm and embrace the limits and imperfections of the world."[62]

As I mentioned earlier, we need to find a way of living with imperfection, which means that we need a way of coming to see life in the world as good and worth affirming despite the ill, that is, we need a way of coming to be at home in the world amidst imperfection. This is precisely what the practice of the Sabbath is supposed to help us achieve. A Sabbath-orientation points us toward the Sabbath experience as the height of our earthly existence, which involves our coming through contemplation to an affirmation of the world and our life within it as "very good," and thereby becoming at home in the world. As Rabbi Abraham Joshua Heschel puts it: "The Sabbath…is more than an armistice, more than an interlude; it is a profound conscious harmony of man and the world, a sympathy for all things."[63] The Sabbath is "the climax of living" (14). We see something similar expressed by Josef Pieper in his account of leisure, which he explicitly connects with the Sabbath experience; he writes:

> Leisure is the disposition of receptive understanding, contemplative beholding, and immersion—in the real…. [It] is the condition of considering things in a celebrating spirit…. Leisure is only possible in the assumption that man is not only in harmony with himself…, but also that he is in agreement with the world and its meaning. Leisure lives on affirmation…. And as it is written in the Scriptures, God saw, when "he rested from all the works that He had made" that everything was good, very good (Genesis 1, 31), just so the leisure of man includes within itself a celebratory, approving, lingering gaze of the inner eye on the reality of creation.[64]

---

[62] Michael J. Sandel, "Mastery and Hubris in Judaism: What's Wrong with Playing God?," in *Public Philosophy* (Cambridge, MA: Harvard University Press, 2005), 205.

[63] Abraham Joshua Heschel, *The Sabbath: Its Meaning for Modern Man* (New York: Farrar, Straus and Giroux, 1951), 31. In other words, the Sabbath is not simply what William James calls a "moral holiday" (*Pragmatism* [Indianapolis, IN: Hackett, 1981 (1907)], 36–8).

[64] Josef Pieper, *Leisure: The Basis of Culture*, trans. Gerard Malsbary (South Bend, IN: St. Augustine's Press, 1998 [1948]), 31, 33.

We see both Heschel and Pieper speak of harmony with the world, which is another way of talking about being at home in the world. In fact, Pieper goes on to say: "no more intensive harmony with the world can be thought of than that of 'Praise of God,' the worship of the Creator of this world" (50). The deepest root from which leisure draws its sustenance, he maintains, is "worshipful celebration" (55). However, even if one is not a theist, one can recognize the importance of the harmony that is sought here. Indeed, in Chapter 1 we saw Cohen speak of seeking to be "at peace with the world" and therefore with oneself, and he also said that "*true* religion celebrates life, and the world, and looks for the good in everything," and, as we have seen here, this is precisely what one does when rightly observing the Sabbath.

The celebration of human finitude and the affirmation and embrace of the limits and imperfections of the world: this encapsulates well the spirit in which I have sought to articulate and defend the virtues of limits. As we saw Berry put it, "to understand ourselves as limited creatures in a limited world" is not a condemnation, but rather "it returns us to our real condition and to our human heritage," and it saves us from the illusory and harmful aspiration toward a limitless condition. I can do no better than to end by recalling these remarks: "We must have limits or we will cease to exist as humans; perhaps we will cease to exist, period.... [Our] human and earthly limits, properly understood, are not confinements but rather inducements to formal elaboration and elegance, to *fullness* of relationship and meaning."

Works Cited

Anderson, Elizabeth, "What Is the Point of Equality?," *Ethics* 109:2 (1999): 287–337.

Anderson, Elizabeth, "The Fundamental Disagreement between Luck Egalitarians and Relational Egalitarians," *Canadian Journal of Philosophy*, Supplemental Volume 36 (2010): 1–23.

Angier, Tom, "Aristotle on Work," *Revue Internationale de Philosophie* 278:4 (2016): 435–49.

Anscombe, G. E. M., "Modern Moral Philosophy," in *Ethics, Religion and Politics: Collected Philosophical Papers*, vol. III (Minneapolis, MN: University of Minnesota Press, 1981), 26–42.

Anscombe, G. E. M., "Murder and the Morality of Euthanasia," *Human Life, Action and Ethics: Essays by G. E. M. Anscombe*, ed. Mary Geach and Luke Gormally (Charlottesville, VA: Imprint Academic, 2005), 261–78.

Anscombe, G. E. M., "Contraception and Chastity," in *Faith in a Hard Ground: Essays on Religion, Philosophy and Ethics by G. E. M. Anscombe*, ed. Mary Geach and Luke Gormally (Charlottesville, VA: Imprint Academic, 2008), 170–91.

Aquinas, St. Thomas, *Summa Theologica*, 5 vols., trans. Fathers of the English Dominican Province (Notre Dame, IN: Ave Maria Press, 1948 [1266–73]).

Aristotle, *Politics*, trans. C. D. C. Reeve (Indianapolis, IN: Hackett, 1998 [*c*.325 BC]).

Aristotle, *Nicomachean Ethics*, 2nd edn, trans. Terence Irwin (Indianapolis, IN: Hackett, 1999 [*c*.325 BC]).

Aristotle, *Nicomachean Ethics*, trans. C. D. C. Reeve (Indianapolis, IN: Hackett, 2014 [*c*.325 BC]).

Ballantyne, Nathan, *Knowing Our Limits* (Oxford: Oxford University Press, 2019).

Baumeister, Roy and Mark Leary, "The Need to Belong: Desire for Interpersonal Attachments as a Fundamental Human Motivation," *Psychological Bulletin* 117:3 (1995): 497–529.

Becker, Gary S., *The Economic Approach to Human Behavior* (Chicago: University of Chicago Press, 1976).

Benedict, St., *The Rule of St. Benedict in English*, ed. Timothy Fry, OSB (Collegeville, MN: Liturgical Press, 1981 [*c*.535–40]).

Berkowitz, Peter, *Constitutional Conservatism: Liberty, Self-Government, and Political Moderation* (Stanford, CA: Hoover Institution Press, 2013).

Berry, Wendell, *The Unsettling of America: Culture & Agriculture* (New York: Avon Books, 1977).

Berry, Wendell, *Home Economics* (New York: North Point Press, 1987).

Berry, Wendell, "Health Is Membership," in *The Art of the Commonplace: The Agrarian Essays of Wendell Berry* (Washington DC: Shoemaker & Hoard, 2002), 144–58.

Berry, Wendell, "Faustian Economics: Hell Hath No Limits," *Harper's Magazine* (May 2008): 35–42.

Berry, Wendell, *It All Turns on Affection: The Jefferson Lecture & Other Essays* (Berkeley, CA: Counterpoint, 2012).

Berry, Wendell, "The Thought of Limits in a Prodigal Age," in *The Art of Loading Brush: New Agrarian Writings* (Berkeley, CA: Counterpoint, 2017), 19–56.

Blake, William, *Jerusalem: The Emanation of the Giant Albion* (Princeton, NJ: Princeton University Press, 1998 [1805]).

Bostrom, Nick, "The Fable of the Dragon Tyrant," *Journal of Medical Ethics* 31:5 (2005): 273–7.

Brague, Rémi, *The Kingdom of Man: Genesis and Failure of the Modern Project*, trans. Paul Seton (Notre Dame, IN: University of Notre Dame Press, 2018).

Brennan, Jason, *Why Not Capitalism?* (New York: Routledge, 2014).

Brennan, Jason, *Why It's OK to Want to Be Rich* (New York: Routledge, 2021).

Burke, Edmund, *Reflections on the Revolution in France* (1790), in *Select Works of Edmund Burke*, vol. II (Indianapolis, IN: Liberty Fund, 1999).

Buss, Sarah, "Appearing Respectful: The Moral Significance of Manners," *Ethics* 109 (1999): 795–826.

Calhoun, Cheshire, "The Virtue of Civility," *Philosophy and Public Affairs* 29:3 (2000): 251–75.

Calhoun, Cheshire, "On Being Content with Imperfection," *Ethics* 127:2 (2017): 327–52.

Camus, Albert, "The Human Crisis" (1946), trans. Lionel Abel, in *Civil Liberties and the Arts: Selections from* Twice a Year, *1938–48*, ed. William Wasserstrom (Syracuse, NY: Syracuse University Press, 1964), 240–54

Chesterton, G. K., *Heretics* (1905), in *The Collected Works of G. K. Chesterton*, vol. I (San Francisco, CA: Ignatius Press, 1986).

Chesterton, G. K., *Orthodoxy* (1908), in *The Collected Works of G. K. Chesterton*, vol. I (San Francisco, CA: Ignatius Press, 1986).

Chesterton, G. K., "The Patriotic Idea" (1904), in *The Collected Works of G. K. Chesterton*, vol. XX (San Francisco, CA: Ignatius Press, 2001), 595–622.

Chesterton, G. K., *The Autobiography of G. K. Chesterton* (San Francisco, CA: Ignatius Press, 2006 [1936]).

Chesterton, G. K., "If I Had Only One Sermon to Preach" (1950), in *In Defense of Sanity: The Best Essays of G. K. Chesterton* (San Francisco, CA: Ignatius Press, 2011), 341–50.

Chesterton, G. K., *The Defendant* (Mineola, NY: Dover [2012 (1902)]).

Clarke, Steve, Julian Savulescu, C. A. J. Coady, Alberto Giubilini, and Sagar Sanyal (eds.), *The Ethics of Human Enhancement: Understanding the Debate* (Oxford: Oxford University Press, 2016).

Cohen, Andrew I., "Famine Relief and Human Virtue," in *Contemporary Debates in Applied Ethics*, 2nd edn, ed. Andrew I. Cohen and Christopher Heath Wellman (Malden, MA: Blackwell, 2014), 431–46.

Cohen, G. A., "On the Currency of Egalitarian Justice," *Ethics* 99:4 (1989): 906–44.

Cohen, G. A., *If You're an Egalitarian, How Come You're So Rich?* (Cambridge, MA: Harvard University Press, 2000).

Cohen, G. A., *Rescuing Justice and Equality* (Cambridge, MA: Harvard University Press, 2008).

Cohen, G. A., *Why Not Socialism?* (Princeton, NJ: Princeton University Press, 2009).

Cohen, G. A., *Finding Oneself in the Other* (Princeton, NJ: Princeton University Press, 2013).

Confucius, *Analects*, trans. Edward Slingerland (Indianapolis, IN: Hackett, 2003 [*c*.450 BC]).

Cooper, David E., *The Mystery of Things: Humanism, Humility, and Mystery* (Oxford: Oxford University Press, 2002).

Cottingham, John, "Partiality and the Virtues," in *How Should One Live?: Essays on the Virtues*, ed. Roger Crisp (Oxford: Clarendon Press, 1996), 57–76.

Cottingham, John, *On the Meaning of Life* (New York: Routledge, 2003).

Cottingham, John, *Cartesian Reflections: Essays on Descartes's Philosophy* (Oxford: Oxford University Press, 2008).

Craiutu, Aurelian, *Faces of Moderation: The Art of Balance in an Age of Extremes* (Philadelphia, PA: University of Pennsylvania Press, 2017).

Deneen, Patrick J., *Why Liberalism Failed* (New Haven, CT: Yale University Press, 2018).

Diamond, Cora, "Eating Meat and Eating People," *Philosophy* 53:206 (1978): 465–79.

Diamond, Cora, "Ethics, Imagination, and the Method of Wittgenstein's *Tractatus*," in *The New Wittgenstein*, ed. Alice Crary and Rupert Read (New York: Routledge, 2000), 149–73.

Diamond, Cora, "The Problem of Impiety," in *Spirituality and the Good Life: Philosophical Approaches*, ed. David McPherson (Cambridge: Cambridge University Press, 2017), 29–46.

Dostoevsky, Fyodor, *The Brothers Karamazov*, trans. Constance Garnett, rev. Ralph E. Matlaw (New York: Norton, 1976 [1880]).

Dostoevsky, Fyodor, *The Brothers Karamazov*, trans. Richard Pevear and Larissa Volokhonsky (New York: Everyman's Library, 1990 [1880]).

Dworkin, Ronald, *Sovereign Virtue: The Theory and Practice of Equality* (Cambridge, MA: Harvard University Press, 2000).

Elias, Norbert, *The Civilizing Process, vol. 1: The History of Manners*, trans. Edmund Jephcott (New York: Pantheon, 1978).

Ellis, Fiona, "Desire and the Spiritual Life," in *Spirituality and the Good Life: Philosophical Approaches*, ed. David McPherson (Cambridge: Cambridge University Press, 2017), 84–100.

Elshtain, Jean Bethke, *Augustine and the Limits of Politics* (Notre Dame, IN: University of Notre Dame Press, 1995).

Elshtain, Jean Bethke, *Sovereignty: God, State, and Self* (New York: Basic Books, 2008).

Fingarette, Herbert, *Confucius: The Secular as Sacred* (Long Grove, IL: Waveland Press, 1972).

Fleischacker, Samuel, "The Jewish Sabbath as a Spiritual Practice," in *Spirituality and the Good Life: Philosophical Approaches*, ed. David McPherson (Cambridge: Cambridge University Press, 2017), 117–35.

Fourie, Carina and Annette Rid (eds.), *What Is Enough?: Sufficiency, Justice, and Health* (Oxford: Oxford University Press, 2017).

Frankfurt, Harry, "Equality as a Moral Ideal," *Ethics* 98:1 (1987): 21–43.

Geach, Peter, *God and the Soul* (London: Routledge, 1978).

Genovese, Eugene D., *The Southern Tradition: The Achievements and Limitations of an American Conservatism* (Cambridge, MA: Harvard University Press, 1994).

Goodhart, David, *The Road to Somewhere: The Populist Revolt and the Future of Politics* (London: Hurst, 2017).

Grimm, Jacob and Wilhelm Grimm, *The Complete Grimm's Fairy Tales*, trans. Margaret Hunt (New York: Race Point, 2013 [1812]).

Hamilton, Alexander, John Jay, and James Madison, *The Federalist*, Gideon edition (Indianapolis, IN: Liberty Fund, 2001 [1787–8]).

Hampshire, Stuart, *Morality and Conflict* (Cambridge, MA: Harvard University Press, 1983).

Hanley, Ryan Patrick, *Adam Smith and the Character of Virtue* (Cambridge: Cambridge University Press, 2009).

Hansen, William, *Classical Mythology: A Guide to the Mythical World of the Greeks and Romans*, 2nd edn (Oxford: Oxford University Press, 2020).

Hartman, David, *A Living Covenant: The Innovative Spirit in Traditional Judaism* (New York: Free Press, 1985).

Hartman, David, *A Heart of Many Rooms: Celebrating the Many Voices within Judaism* (Woodstock, VT: Jewish Lights Publishing, 1999).

Hauskeller, Michael, "Human Enhancement and the Giftedness of Life," *Philosophical Papers* 40:1 (2011): 55–79.

Haybron, Dan, *Happiness: A Very Short Introduction* (Oxford: Oxford University Press, 2013).

Hegel, G. W. F., *The Philosophy of Right*, trans. T. M. Knox (Oxford: Oxford University Press, 1967 [1821]).

Hegel, G. W. F., *Hegel's Logic: Being Part One of the* Encyclopedia of the Philosophical Science, trans. William Wallace (Oxford: Oxford University Press, 1975 [1830]).

Hegel, G. W. F., *Aesthetics: Lectures on the Fine Arts*, trans. T. M. Knox (Oxford: Oxford University Press, 1975 [1835]).

Heschel, Abraham Joshua, *The Sabbath: Its Meaning for Modern Man* (New York: Farrar, Straus and Giroux, 1951).

Hirschfield, Mary L., *Aquinas and the Market: Toward a Humane Economy* (Cambridge, MA: Harvard University Press, 2018).

Hume, David, *An Enquiry Concerning the Principles of Morals* (1751), in *Enquiries Concerning Human Understanding and Concerning the Principles of Morals*, 3rd edn (Oxford: Clarendon Press, 1975).

Hume, David, *A Treatise of Human Nature*, 2nd edn (Oxford: Clarendon Press, 1978 [1739]).

Hursthouse, Rosalind, *On Virtue Ethics* (Oxford: Oxford University Press, 1999).

James, William, *Pragmatism* (Indianapolis, IN: Hackett, 1981 [1907]).

John Paul II, Pope, *Laborem Exercens* (1981), reprinted in *Catholic Social Thought: The Documentary Heritage*, ed. David J. O'Brien and Thomas A. Shannon (Maryknoll, NY: Orbis Books, 1992).

John Paul II, Pope, *Veritatis Splendor* (1993), http://www.vatican.va/content/john-paul-ii/en/encyclicals/documents/hf_jp-ii_enc_06081993_veritatis-splendor.html, accessed July 27, 2021.

Jollimore, Troy, *On Loyalty* (New York: Routledge, 2013).

Kahane, Guy, "Mastery without Mystery: Why There Is No Promethean Sin in Enhancement," *Journal of Applied Philosophy* 28:4 (2011): 355–68.

Kass, Leon R., "The Wisdom of Repugnance," *The New Republic* (June 2, 1997), 17–26.

Kass, Leon R., *The Hungry Soul: Eating and the Perfecting of Our Nature* (Chicago: University of Chicago Press, 1999 [1994]).

Kass, Leon R., "*L'Chaim* and Its Limits: Why Not Immorality?," in *Life, Liberty, and the Defense of Dignity: The Challenge for Bioethics* (San Francisco, CA: Encounter, 2002), 257–74.

Kass, Leon R., *The Beginning of Wisdom: Reading Genesis* (Chicago: University of Chicago Press, 2003).

Kekes, John, *The Illusions of Egalitarianism* (Ithaca, NY: Cornell University Press, 2003).

Keller, Simon, *The Limits of Loyalty* (Cambridge: Cambridge University Press, 2007).

Keynes, John Maynard, "Economic Possibilities for Our Grandchildren" (1930), in *The Collected Writings of John Maynard Keynes*, vol. IX: *Essays in Persuasion* (Cambridge: Cambridge University Press, 2015), 321–32.

Kleinig, John, "Loyalty," *Stanford Encyclopedia of Philosophy*, 2013, http://plato.stanford.edu/entries/loyalty/, accessed July 28, 2021.

Kleinig, John, *On Loyalty and Loyalties: The Contours of a Problematic Virtue* (Oxford: Oxford University Press, 2014).

Kolnai, Aurel, *The Utopian Mind and Other Papers*, ed. Francis Dunlap (London: Athlone, 1995).

Kolnai, Aurel, "Privilege and Liberty" (1949), in *Privilege and Liberty and Other Essays in Political Philosophy* (Lanham, MD: Lexington, 1999), 19–61.

Kolozi, Peter, *Conservatives against Capitalism: From the Industrial Revolution to Globalization* (New York: Columbia University Press, 2017).

Kukathas, Chandran, "The Case for Open Immigration," in *Contemporary Debates in Applied Ethics*, 2nd edn, ed. Andrew I. Cohen and Christopher Heath Wellman (Malden, MA: Blackwell, 2014), 376–88.

Lasch, Christopher, *The True and Only Heaven: Progress and Its Critics* (New York: Norton, 1991).

Lasch, Christopher, *The Revolt of the Elites and the Betrayal of Democracy* (New York: Norton, 1996).

Leiss, William, *The Domination of Nature* (Boston, MA: Beacon, 1972).

Lind, Michael, *The New Class War: Saving Democracy from the Managerial Elite* (New York: Portfolio, 2020).

McCarthy, Cormac, *No Country for Old Men* (New York: Vintage, 2005).

McClure, Jane Michele, OSB, Sr. "Benedictine Way of Life" (2015), http://stellamaris. nsw.edu.au/wp-content/uploads/2015/06/benedictine-way-of-life.pdf, accessed July 28, 2021.

McPherson, David, "Nietzsche, Cosmodicy, and the Saintly Ideal," *Philosophy* 91:1 (2016): 39–67.

McPherson, David, "Existential Conservatism," *Philosophy* 94:3 (2019): 383–407.

McPherson, David, *Virtue and Meaning: A Neo-Aristotelian Perspective* (Cambridge: Cambridge University Press, 2020).

MacIntyre, Alasdair, *Dependent Rational Animals: Why Human Beings Need the Virtues* (Chicago: Open Court, 1999).

Mandeville, Bernard, *The Fable of the Bees: or, Private Vices, Public Benefits* (1714), in *The Fable of the Bees and Other Writings* (Indianapolis, IN: Hackett, 1997).

Marx, Karl and Friedrich Engels, *The Communist Manifesto* (1848), in *Karl Marx: Selected Writings*, ed. David McLellan, 2nd edn (Oxford: Oxford University Press, 2000).

Marx, Karl, "Economic and Philosophical Manuscripts" (1844), in *Karl Marx: Selected Writings*, ed. David McLellan, 2nd ed. (Oxford: Oxford University Press, 2000), 83–121.

May, Simon, "Why Nietzsche Is Still in the Morality Game," in *Nietzsche's* On the Genealogy of Morality: *A Critical Guide*, ed. Simon May (Cambridge: Cambridge University Press, 2011), 78–100.

Meilaender, Gilbert, *Bioethics: A Primer for Christians*, 2nd edn (Grand Rapids, MI: Eerdmans, 2004).

Mencius, *Mencius*, trans. Bryan W. Van Norden, in *Readings in Classical Chinese Philosophy*, ed. Philip J. Ivanhoe and Bryan W. Van Norden (Indianapolis, IN: Hackett, 2001).

Merton, Thomas, *The Sign of Jonas* (New York: Harvest Book, 1953).

Midgley, Mary, "On Not Being Afraid of Natural Sex Differences," in Morwenna Griffiths and Margaret Whitford (eds.), *Feminist Perspectives in Philosophy* (London: Macmillan Press, 1988), 29–41.

Mill, John Stuart, *Utilitarianism* (Indianapolis, IN: Hackett, 2001 [1861]).

Miller, David, *Justice for Earthlings: Essays in Political Philosophy* (Cambridge: Cambridge University Press, 2013).

Miller, David, "Immigration: The Case for Limits," in *Contemporary Debates in Applied Ethics*, 2nd edn, ed. Andrew I. Cohen and Christopher Heath Wellman (Malden, MA: Blackwell, 2014), 363–75.

Mitchell, Mark T., *The Politics of Gratitude: Scale, Place & Community in a Global Age* (Washington DC: Potomac Books, 2012).

Mitchell, Mark T., "Making Places: The Cosmopolitan Temptation," in *Why Place Matters: Geography, Identity, and Civic Life in Modern America*, ed. Wilfred M. McClay and Ted V. McAllister (New York: Encounter Books, 2014), 84–101.

Murdoch, Iris, "On 'God' and 'Good,'" in *Existentialists and Mystics: Writings on Philosophy and Literature*, ed. Peter Conradi (New York: Penguin, 1997), 337–62.

Nagel, Thomas, *Equality and Partiality* (Oxford: Oxford University Press, 1991).

Nietzsche, Friedrich, *The Antichrist* (1888), trans. Walter Kaufmann, in *The Portable Nietzsche* (New York: Penguin, 1954).

Nietzsche, Friedrich, *The Gay Science*, trans. Walter Kaufmann (New York: Vintage Books, 1974 [1882/1887]).

Nietzsche, Friedrich, *Thus Spoke Zarathustra*, trans. Walter Kaufmann (New York: Modern Library, 1995 [1883–5]).

Nietzsche, Friedrich, *The Birth of Tragedy* (1872), trans. Walter Kaufmann, in *Basic Writings of Nietzsche* (New York: Modern Library, 2000).

Nozick, Robert, *Anarchy, State, and Utopia* (New York: Basic Books, 1974).

Nozick, Robert, *Philosophical Explanations* (Cambridge, MA: Belknap Press of Harvard University, 1981).

Nussbaum, Martha C., "Transcending Humanity," in *Love's Knowledge: Essays on Philosophy and Literature* (Oxford: Oxford University Press, 1990), 365–91.

Nussbaum, Martha C., "Patriotism and Cosmopolitanism," in *For Love of Country?*, ed. Joshua Cohen (Boston, MA: Beacon Press, 1996), 3–17.

Nussbaum, Martha C., *The Cosmopolitan Tradition: A Noble But Flawed Ideal* (Cambridge, MA: Belknap Press of Harvard University Press, 2019).

O'Connor, Flannery, *Collected Works* (New York: Library of America, 1988).

Olberding, Amy, "Etiquette: A Confucian Contribution to Moral Philosophy," *Ethics* 126:2 (2016): 422–46.

Parens, Erik, *Shaping Our Selves: On Technology, Flourishing, and a Habit of Thinking* (Oxford: Oxford University Press, 2015).

Parens, Erik and Josephine Johnston (eds.), *Human Flourishing in an Age of Gene Editing* (Oxford: Oxford University Press, 2019).

Pieper, Josef, *Leisure: The Basis of Culture*, trans. Gerard Malsbary (South Bend, IN: St. Augustine's Press, 1998 [1948]).

Pieper, Josef, *In Tune with the World: A Theory of Festivity*, trans. Richard and Clare Winston (South Bend, IN: St. Augustine's Press, 1999 [1963]).

Platts, Mark, "Moral Reality and the End of Desire," in *Reference, Truth and Reality*, ed. Mark Platts (London: Routledge & Kegan Paul, 1980), 69–82.

Polacco, Patricia, *Luba and the Wren* (New York: Philomel, 1999).

Pugh, Jonathan, Guy Kahane, and Julian Savulescu, "Cohen's Conservatism and Human Enhancement," *The Journal of Ethics* 17:4 (2013): 331–54.

Quinton, Anthony, *The Politics of Imperfection* (London: Faber and Faber, 1978).

Rawls, John, *A Theory of Justice*, revised edn (Cambridge, MA: Harvard University Press, 1999 [1971]).

Rawls, John, *Political Liberalism*, expanded edn (New York: Columbia University Press, 2005 [1993]).

Reginster, Bernard, *The Affirmation of Life: Nietzsche on Overcoming Nihilism* (Cambridge, MA: Harvard University Press, 2006).

Roberts, Robert C., "The Blessings of Gratitude: A Conceptual Analysis," in *The Psychology of Gratitude*, ed. Robert A. Emmons and Michael E. McCullough (Oxford: Oxford University Press, 2004), 58–78.

Roberts, Robert C., "Temperance," in *Virtues and Their Vices*, ed. Kevin Timpe and Craig Boyd (Oxford: Oxford University Press, 2014), 93–111.

Roberts, Robert C., (with Cleveland, W. Scott), "Humility from a Philosophical Point of View," in *Handbook of Humility: Theory, Research, and Applications*, ed. Everett L. Worthington et al. (New York: Routledge, 2017), 33–46.

Robinson, Marilynne, *Gilead* (New York: Farrar, Straus and Giroux, 2004).

Röpke, Wilhelm, *A Humane Economy: The Social Framework of the Free Market* (Chicago: Regnery, 1960).

Sandel, Michael J., "The Procedural Republic and the Unencumbered Self," *Political Theory* 12:1 (1984): 81–96.

Sandel, Michael J., *Democracy's Discontent: America in Search of a Public Philosophy* (Cambridge, MA: Belknap Press of Harvard University, 1996).

Sandel, Michael J., "Mastery and Hubris in Judaism: What's Wrong with Playing God?," in *Public Philosophy* (Cambridge, MA: Harvard University Press, 2005), 196–210.

Sandel, Michael J., *The Case against Perfection: Ethics in an Age of Genetic Engineering* (Cambridge, MA: Belknap Press of Harvard University Press, 2007).

Sandel, Michael J., *What Money Can't Buy: The Moral Limits of Markets* (New York: Farrar, Straus and Giroux, 2012).

Sandel, Michael J., *The Tyranny of Merit: What's Become of the Common Good?* (New York: Farrar, Straus and Giroux, 2020).

Savulescu, Julian and Nick Bostrom (eds.), *Human Enhancement* (Oxford: Oxford University Press, 2009).

Schumaker, E. F., *Small Is Beautiful: Economics as if People Mattered* (New York: Harper, 1973).

Scruton, Roger, *Sexual Desire: A Philosophical Investigation* (New York: Continuum, 2006 [1986]).

Scruton, Roger, "Meaningful Marriage," in *A Political Philosophy: Arguments for Conservatism* (London: Bloomsbury, 2006), 81–102.

Scruton, Roger, *The Uses of Pessimism: And the Dangers of False Hope* (Oxford: Oxford University Press, 2010).

Scruton, Roger, *How to Think Seriously about the Planet: The Case for an Environmental Conservatism* (Oxford: Oxford University Press, 2012).

Scruton, Roger, "Real Men Have Manners," in *The Philosophy of Food*, ed. David M. Kaplan (Berkeley, CA: University of California Press, 2012), 24–32.

Scruton, Roger, *How to Be a Conservative* (London: Bloomsbury, 2014).

Sher, George, *Equality for Inegalitarians* (Cambridge: Cambridge University Press, 2014).

Shields, Philip R., *Logic and Sin in the Writings of Ludwig Wittgenstein* (Chicago: University of Chicago Press, 1993).

Shiffman, Mark, "The Rediscovery of *Oikonomia*," in *The Humane Vision of Wendell Berry*, iBook edn, ed. Mark T. Mitchell and Nathan Schlueter (Wilmington, DE: ISI Books, 2011), 349–91.

Singer, Peter, "Famine, Affluence, and Morality," *Philosophy & Public Affairs* 1:3 (1972): 229–43.

Skidelsky, Robert and Edward Skidelsky, *How Much Is Enough?: Money and the Good Life* (New York: Other Press, 2012), 3.

Smith, Adam, *The Theory of Moral Sentiments* (Indianapolis, IN: Liberty Fund, 1985 [1759]).

Smith, Adam, *An Inquiry into the Nature and Causes of the Wealth of Nations*, vol. I (Indianapolis, IN: Liberty Fund, 2009 [1776]).

Snow, Nancy E., "Humility," *Journal of Value Inquiry* 29:2 (1995): 203–16.

Stegner, Wallace, *Where the Bluebird Sings to the Lemonade Springs: Living and Writing in the West* (New York: Random House, 1992).

Stohr, Karen, *On Manners* (New York: Routledge, 2011).

Taylor, Charles, *Sources of the Self: The Making of Modern Identity* (Cambridge, MA: Harvard University Press, 1989).

Taylor, Charles, *The Language Animal: The Full Shape of the Human Linguistic Capacity* (Cambridge, MA: Belknap Press of Harvard University, 2016).

Toner, Christopher, "Home and Our Need for It," *Journal of Philosophical Research* 44 (2019): 251–72.

Tocqueville, Alexis de, *Democracy in America*, trans. Harvey C. Mansfield and Delba Winthrop (Chicago: University of Chicago Press, 2002 [1835/1840]).

Waldron, Jeremy, "Who is My Neighbor?: Humanity and Proximity," *The Monist* 86:3 (2003): 333–54.

Walzer, Michael, *Spheres of Justice: A Defense of Pluralism and Equality* (New York: Basic Books, 1983).

Weber, Michael and Kevin Vallier (eds.), *Political Utopias: Contemporary Debates* (Oxford: Oxford University Press, 2017).

Weckert, John, "Playing God: What is the Problem?," in Steve Clarke, Julian Savulescu, C. A. J. Coady, Alberto Giubilini, and Sagar Sanyal (eds.), *The Ethics of Human Enhancement: Understanding the Debate* (Oxford: Oxford University Press, 2016), 87–99.

Weil, Simone, *The Need for Roots* (New York: Routledge, 1952).

Westacott, Emrys, *The Wisdom of Frugality: Why Less Is More—More or Less* (Princeton, NJ: Princeton University Press, 2016).

Whitcomb, Dennis, Heather Battaly, Jason Baehr, and Daniel Howard-Snyder, "Intellectual Humility: Owning Our Limitations," *Philosophy and Phenomenological Research* 94:3 (2017): 509–39.

Wiggins, David, "Claims of Need," in *Needs, Values, Truth*, 3rd edn (Oxford: Clarendon Press, 1998 [1987]), 1–57.

Wiggins, David, "Nature, Respect for Nature, and the Human Scale of Values," *Proceedings of the Aristotelian Society* 100 (2000): 1–32.

Wiggins, David, *Sameness and Substance Renewed* (Cambridge: Cambridge University Press, 2001).

Wiggins, David, *Ethics: Twelve Lectures on the Philosophy of Morality* (Cambridge, MA: Harvard University Press, 2006).

Wiggins, David, "Work, Its Moral Meaning or Import," *Philosophy* 89:3 (2014): 477–82.

Williams, Bernard, "A Critique of Utilitarianism," in *Utilitarianism: For and Against,* coauthored with J. J. C. Smart (Cambridge: Cambridge University Press, 1973), 75–150.

Williams, Bernard, *Moral Luck* (Cambridge: Cambridge University Press, 1981).

Williams, Bernard, *Ethics and the Limits of Philosophy* (Cambridge, MA: Harvard University Press, 1985).

Williams, Bernard, "Must a Concern for the Environment Be Centered on Human Beings?," in *Making Sense of Humanity: And Other Philosophical Papers* (Cambridge: Cambridge University Press, 1995), 233–40.

Williams, Bernard, *Truth and Truthfulness* (Princeton, NJ: Princeton University Press, 2002).

Winch, Peter, "Who is My Neighbour?," in *Trying to Make Sense.* (Oxford: Blackwell, 1987), 154–66.

Wirzba, Norma, *Living the Sabbath: Discovering the Rhythms of Rest and Delight* (Grand Rapids, MI: Brazos Press, 2006).

Wirzba, Norma, *Food and Faith: A Theology of Eating* (Cambridge: Cambridge University Press, 2011).

Wittgenstein, Ludwig, *Notebooks, 1914–1916*, 2nd edn, trans. G. E. M. Anscombe (Chicago: University of Chicago Press, 1961).

Woodruff, Paul, *Reverence: Renewing a Forgotten Virtue*, 2nd edn (Oxford: Oxford University Press, 2014).

Xunzi, *Xunzi*, trans. Eric L. Hutton, in *Readings in Classical Chinese Philosophy*, ed. Philip J. Ivanhoe and Bryan W. Van Norden (Indianapolis, IN: Hackett, 2001).

Index

For the benefit of digital users, table entries that span two pages (e.g., 52–53) may, on occasion, appear on only one of those pages.

absolute prohibitions 46, 62–71
Aristotle on 63
character-centered approach 63–4, 67–8
divine law approach 63–4, 68–71
intrinsic reasons for absolute prohibitions 69
moral identity approach 65–7
preventing moral anarchy approach 63–6
accepting-appreciating stance (accepting the given) 5, 17, 21–2, 34, 47–8, 81, 126
see also choosing-controlling stance (seeking improvement); the given; loyalty to the given
accepting some things as given 17
case for regarding as primary 15, 18–20, 23
choosing-controlling stance dominance over 6–7, 12, 14–20, 25
and intrinsic values 17–20, 26–7
and loyalty to the given 44–5
Nietzsche appearing to adopt 25–7
no place given to 16–17
relationship with choosing-controlling stance 15–16, 18–20
spiritual requirements 17
and sufficientarianism 108
yes-saying 25–7, 34–5
acquisitiveness 126–7, 130–1, 133–4
see also capitalism; greed, vice of; money; wealth
limits on 132–3, 144–5

Adam and Eve 59, 148
adultery 70–1
Aeschylus
Prometheus Bound 5–6
agent-centered approach, absolute prohibitions 69–70
alienation 80–1, 126, 145, 148, 151–2
existential 22
Anderson, Elizabeth 100–1, 104–5
Anscombe, Elizabeth 67–71
"Modern Moral Philosophy" 67–9
appreciation
aim of life 37
and contentment 35–6, 40
of the given 8, 105–7
and gratitude 35, 38–9
and intrinsic values 137
of life as a gift 24, 27–30, 142–3
of limits 159–60
patriotism 87
Aquinas, St. Thomas 29–30, 92–3
Aristotle 28–9, 48, 50, 68, 70, 77–8, 88, 135, 148–50, 154–5
on absolute prohibitions 63
on ethics 58, 131–2
Nicomachean Ethics 48, 71–2
on *eudaimonia* (good life) 131–2, 136–7
on friendship 82–3
on justice 114–16
on moderation 120
on *oikonomia* (household management) 126, 144–5
on *pleonexia* (overreaching in benefits for ourselves) 114–15, 131–2

Aristotle (*cont.*)
  Politics 46–7, 144
  on private property 145–6
  on temperance 57–60
  on virtues of character 62–3
assistance, duties of 71–83
  positive nature of 71
autonomy principle 14–15, 27–8
avarice 126–7

basic human goods 132–4
beauty 26–7
belonging, need for sense of 88,
  92–3, 98
beneficence principle 14–15, 27–8
Berry, Wendell 147, 150–5, 161
  "Faustian Economics" 151–2
  "It All Turns on Affection" 153
  "The Thought of Limits in a Prodigal
    Age" 153
  The Unsettling of America 146–7
Bible *see* Genesis, Hebrew Bible; Gospel
  of Matthew; Parable of the Good
  Samaritan; "Sermon on the
  Mount"
Blake, William 75, 153
borders, open 94
Bostrom, Nick 12–13, 18–20
Brague, Rémi 6–7
Brennan, Jason 136–8, 140
  Why It's OK to Want to Be Rich
    135–6
  Why Not Capitalism? 135
brute luck 100–2
Buchanan, Allen 18–20
Burke, Edmund 119
Buss, Sarah 52–4, 56

Calhoun, Cheshire 35, 39
  "On Being Content with
    Imperfection" 35
Camus, Albert 123–4
capitalism
  see also greed, vice of; Marx, Karl;
    money; wealth

  Cohen on 112–13
  economic decentralization 145–6
  as a force for good or bad 126–7,
    130–1
  global 97–8
  and greed 112–13, 129
  ideology 130–1
  industrial 90–1
  and love of money 130–1
  Marx on 130–1, 145–6
  progressive optimism 125
  versus socialism 135–6
  utopian vision 125
  worker exploitation (Marx) 130–1
centralization of political life 89–90
character-centered approach, absolute
  prohibitions 63–4, 67–8
character formation
  see also good manners
  Aristotelian in nature 48
  Confucian in nature 48–9
  good manners, importance of 53–4
  moderation 2–3, 48, 57–62
  moral limits 46–62
  potential savagery of humans 46–7
  restraints on desires 46–9, 62–71
  reverence/respect 48, 51–7
  virtue of character as a "medial
    condition" 62–3
  virtuous actions 48
Chesterton, G. K. 24–5, 37–8, 74, 81,
  88, 93–4
  *The Autobiography of G. K.
    Chesterton* 37
  on loyalty to the given 40–1, 44
  on man as a Great Might-Not-Have-
    Been 37–8
  Orthodoxy 37, 40, 44
  on patriotism 86–7
children *see* genetic engineering of
  children
choosing-controlling stance (seeking
  improvement)
  see also accepting-appreciating stance
    (accepting the given)

character formation 47–8
contentment 34
economic limits 126
making dominant in seeking mastery
        over the given 6–7, 12,
        14–20, 25
Nietzsche on 14
problems with complete
        dominance 16–17
relationship with accepting-
        appreciating stance 15–16,
        18–20
science and technology 12
Christianity 28–9, 73–4
    see also Gospel of Matthew; Jesus
        Christ; New Testament
    Catholic social teaching 148–50
    Sabbath-orientation/practice
        156–7
citizenship and political
        community 84–98, 107
    see also cosmopolitanism; distributive
        justice; immigration;
        neighborliness; patriotism;
        sufficientarian justice
    bond of political community 108–9,
        112, 119–20
    communities of character 96
    protecting fellow citizens from effects
        of immigration 95–6
    self-government, democratic 84,
        89–90, 92, 94
    and cultural self-determination 96
    world citizenship 86
civic education 86
civility 50
Cleveland, W. Scott 29–30
Cohen, A. I. 75–6
Cohen, G. A. 42, 161
    atheism 20–1, 36–7
    on capitalism 112–13
    on contentment and gratitude 36–7,
        39–40, 118
    "On the Currency of Egalitarian
        Justice" 100, 104–5

and egalitarianism 42, 101–2,
        105–7, 111
    existing value, defense of 41–2
    on justice 115–16, 118
    and opposition to the Promethean
        ideal 17–18, 20–3
    on personal valuing 43
    "Rescuing Conservatism" 16–17, 23,
        41, 99–100, 104–5
    on socialist ideal/utopianism 16–17,
        111–18, 129
    utopianism 116–17
    Why Not Socialism? 111
compassion 10–11
Confucius/Confucian way of life 2
    Analects 48–9, 51–2, 54–5, 81–2
    character formation 48–9, 56
    and filial piety 71–2
consequentialism 17, 46, 65–6
    consequentialist ethics 64
conservatism
    existential 16–17, 104–5, 107
    small-c and large-C 16–17, 104–5
contentment
    accepting-appreciating stance 34
    and appreciation 35–6, 40
    choosing-controlling stance 34
    contrasting with greed 134
    economic limits 134–5, 138–9
    and freedom 138–9
    and frugality/minimalism 134–5
    and gratitude 3, 34–40
    as a limiting virtue 3, 141–2
    in New Testament 138–9
    Nietzsche on 34–5
    orientation toward the given
        world 38–9
    and positive freedom 137
    and sufficiency 136–7
contingency 103–4, 107, 142–3
    "metaphysical crusade against"
        99–100, 104–8
    radical 66–7
Cooper, David E. 5–6
cosmodicy problem 20–1, 107

cosmopolitanism 85, 88, 90–3
    "anywheres" and "somewheres"
        84–5, 89, 94
    cosmopolitan education 86
    limitations of 91–2
Cottingham, John 77–8
    On the Meaning of Life 8
courage 57–8
courtesy 51–2
    see also good manners; respect
Craiutu, Aurelian 120–2
creation story, Genesis 22–3, 155–6
creativity 6–7, 9, 15

Dao (Confucian way of life) *see*
        Confucius/Confucian way of life
"death of God" *see* God
decentralization
    economic 145–6, 154–5
    political 89–90, 92–3, 98, 145–6
dehumanization 1, 5, 18–20, 34–5,
        48–9, 151
desires
    see also restraints on desires
    and character formation 46–8
    giving form to 47–8
    and humans considered as potentially
        the best of animals 46–50
    indefinite expansion of 124–5
    insatiability of 10–11, 59–60, 126–8,
        138–9
        for wealth 130–3, 136–7, 141–2
    intrinsic values 18
    and opposition to the Promethean
        ideal 23–5
    satisfaction of 6–7, 34–5, 138–9,
        143–4
    sexual 62, 78–9
Diamond, Cora 30–1, 52–3, 141–2
difference principle 100–1, 103
dignity of human life 18, 27–8
    see also murder
    character formation 53–6
    humility and reverence 28–31, 142–3
    murder violating 69

respect for life 69–70
violation of intrinsic value of human
        life 63
dignity of work 126, 148–50
distributism 145–6
distributive justice 3, 84, 96–7
    see also immigration; self-
        government, democratic;
        sufficientarian justice
    and egalitarianism 99, 101–2,
        107–8, 110
        feasibility problem 110
    and immigration 95–8
    and luck egalitarianism 99
    rights and liberties 98–9
    and spirituality 104–5
    sufficientarian conception of 98–9,
        108–9, 112
divine law approach, absolute
        prohibitions 63–4, 68–71
Dostoevsky, Fyodor
    The Brothers Karamazov 21, 75,
        138–9
duties of assistance *see* assistance,
        duties of
Dworkin, Ronald 7, 10, 105–7
    on genetic engineering 7, 10, 14–15,
        21, 42
    liberal egalitarian outlook 14–15
    Nietzsche compared 14–15
    on playing God 12–14, 21
    and primacy of choosing-controlling
        stance 14
    and threat of nihilism 25, 27–8

economic decentralization 145–6,
        154–5
economic limits 126–61
    contentment 134–5, 138–9
    frugality 134–5
    and good life, living 126, 131–8,
        144–5, 150, 154–5
    greed *see* greed, vice of
    happiness 136–7
    home economics/*oikonomia* 143–55

homo economicus 143–4
minimalism 134–5
positive freedom 137–8
Sabbath-orientation 155–61
education, cosmopolitan 86
egalitarianism
  see also equality; luck egalitarianism
  and distributive justice 99, 101–2,
     107–8, 110
    feasibility problem 110
  ethos 101–2
  moral 99
  relational 99–101
Elias, Norbert
  The Civilizing Process 50
equality
  economic 112
  equality of condition 99, 109
  equality of outcome 99
  and freedom 110, 116–17
  socialist principle 111
ethical individualism 14–15, 27–8
ethic of power (Nietzsche) 10–11, 25–6
ethics
  see also ethic of power (Nietzsche)
  ancient/traditional forms 71–2,
    130–1
  Confucian 71–2
  consequentialist 46, 63–6
  suicide 44–5
  utilitarianism 64
  virtue ethics 1, 67, 135, 137–8
eudaimonia (good life) 131–2, 136–7
  see also good life, living
eugenics 15–16
  see also genetic engineering of
    children
evil 20–2
  greed and love of money 129, 138–9
  and murder 63, 69–70
  and politics 120–1
  and suicide 44
  and tree of knowledge 59, 148
  "unapproachable," in fairy tales
    141–2

existence 56, 79–81, 88, 136–7,
    143–4, 160
  see also non-existence
  dependent nature 30
  gratuitousness of 37, 39
  happiness justifying 12, 34–5
  limits of 21–2, 155–6
  sources 30–1, 54–5, 108
existential conservatism 16–17,
    104–5, 107
existential gratitude 39, 42
existential limits 5–45
  alienation 22
  contentment and gratitude 34–40
  defining 5
  humility and reverence 28–34
  loyalty to the given 40–5
  Promethean ideal (seeking mastery
    over the given) 5–15
    opposition to 15–28
existential stances (orientations
    towards the given) *see*
    accepting-appreciating stance
    (accepting the given);
    choosing-controlling stance
    (seeking improvement)
existential vertigo 8–9, 25
  see also *The Gay Science*, Nietzsche
    ("death of God" passage)
expectation frames 35–6

family relationships *see* filial piety; love;
    loyalty; parents
feasibility problem
  egalitarianism and distribute
    justice 110
  and socialist ideal 113–16
The Federalist
  Federalist No. 1 121–2
  Federalist No. 9 121–2
  Federalist No. 10 122
  Federalist No. 51 122–3
  Federalist No. 55 123
filial piety
  see also piety

filial piety (*cont.*)
  character formation and reverence/
      respect 51–2, 54–6, 60–2
  and Confucian ethics 71–2
  and gratitude 54–5
  humility and reverence 30–1
  and loyalty 80
  parents and children 80
  and patriotism 92–3, 108
Fingarette, Herbert 48–9, 56
**fire, significance in myth of
      Prometheus** 5–6, 17
**"The Fisherman and His Wife"
      (Brothers Grimm fairy
      tale)** 140–2
Fleischacker, Samuel 156–9
Frankfurt, Harry 98–9
freedom
  and contentment 138–9
  and equality 110, 116–17
  and kinship 80–1
  positive 135–40
  principle of 95
  and progressive optimism 125
  voluntarist conception of 90–1
  and wealth 135–6, 140
friendship 71–2, 77–8, 82–3, 114–15
  patriotism *see* patriotism
frugality 134–5

Garden of Eden 59, 148
***The Gay Science*, Nietzsche ("death of
      God" passage)**
  existential vertigo 8–9, 25
  implications 8
  liberation, sense of 9
  limitlessness 9
  "madman" 7–9
gene editing 12–13
generosity 57–8, 131–2
Genesis, Hebrew Bible 129,
      148–50, 160
  creation story 22–3, 155–6
  Garden of Eden 59, 148
  Tower of Babel 6–7

genetic engineering of children
  Dworkin on 7, 10, 14–15, 21, 42
  luck egalitarianism 105–7
  playing God 12–13, 21
  and primacy of choosing-controlling
      stance 15–16
  prioritizing of dominion over
      reverence 30–1
  undermining appreciation of human
      life as a gift 18–20, 27–8
gift, human life perceived as 29–30
  appreciation of 24, 27–30, 142–3
  genetic engineering
      undermining 18–20, 27–8
  and reverence 54–5
the given
  see also Promethean ideal (seeking
      mastery over the given)
  accepting *see* accepting-appreciating
      stance (accepting the given)
  appreciation of 8, 105–7
  loyalty to 40–5
  luck egalitarianism 103–4
  seeking mastery over 6–7, 12,
      14–20, 25
globalization 89–90
God
  see also Nietzsche, Friedrich;
      playing God
  belief in 8
  "death of" see *The Gay Science*,
      Nietzsche ("death of God"
      passage)
  holiness 30–1
  humans in the image of 28–9, 148–50
  Praise of 161
  reverence, worthy of 30–1
godliness 138–9
Goodhart, David 84–5
good life, living 131–5, 144–5
  basic goods necessary for 132–4
  objective understanding 136–8
  and positive freedom 137–9
  and role of economics 126, 131–8,
      144–5, 150, 154–5

self-authorship 135–6
subjective understanding 135–7
virtues as constitutive of 1–2, 51–2,
      133–4, 137–8
good manners 48–54
   importance for character
         formation 53–4, 56
   and respect 52, 56
   teaching of 55–6
   while eating 60–2
Gospel of Matthew 138–9
government, limits of 119–25
gratitude
   see also gift, human life perceived as
   and appreciation 35, 38–9
   and contentment 3, 34–40
   existential 39, 42
   and filial piety 54–5
   giftedness of the world 42
   propositional 39
   Sabbath-orientation/practice 159–60
   stance of 15
greed, vice of 3, 126–34
   see also acquisitiveness; capitalism;
         money; wealth
   and avarice 126–7
   and capitalism 112–13, 129
   conducive to public benefit 127–8
   and economic incentives 129
   "Faustian bargain" 130–1
   reviled human characteristic 126–7
   sacrificing of virtue for economic
         gain 130–1
   seen as a necessary evil 129
   and social status 130–2

Hamilton, Alexander 121–2
Hampshire, Stuart 64–7
Hansen, William 5–6
happiness 12, 136–7
   of "the last man" 34–5
   and money 136–7
   utilitarianism 64
Hartman, Rabbi David 159–60
Hauskeller, Michael 21–2

Hegel, Georg Wilhelm Friedrich
      145–6, 148–52
Heschel, Abraham 160–1
Hobbesian state of nature 65
home economics 143–55
   Aristotle on 144–5
   "boomers" and "stickers"
         distinction 153–4
   views of Berry *see* Berry, Wendell
   views of Hegel *see* Hegel, Georg
         Wilhelm Friedrich
homo economicus 143–4
honor 131–2
hubris (improper pride) 28–30, 33–4,
   meritocratic 142–3
humane localism 88
humanitarianism 10–12
   see also love
   citizenship and political
         community 94, 97–8
humanity 1–3, 5–6, 47–50, 97–8
   see also dehumanization; dignity of
         human life
   abstract 4, 75, 86–7
   being fully human 46
   common 84–5, 88–92, 95, 150
   concrete 4, 75–6
   dignity 53–4
   duty to 74, 91–2
   love for 75–6, 86–7, 91–2
   loyalty to 77
   political community 59–62
   principle of 94–5
   as replaceable 27–8, 42
humanization 59–60
Hume, David 114–15
humility
   see also reverence/respect
   appreciation of dependent nature of
         existence 30
   and contentment 141–2
   improper kind of humility/self-
         abasement 28–9, 33–4
   as the master limiting virtue 2,
         28–9

humility (*cont.*)
   not thinking of oneself as divine
      28–30
   recognition of proper place in scheme
      of things 30, 32
   and reverence 2, 28–34
   self-referential nature of 32
Hursthouse, Rosalind 69–70
Huxley, Aldous
   Brave New World 34–5

ideal theory 84
identity
   see also belonging, need for sense of
   identity-constituting bonds of
      attachment 4, 77–8, 89–90
   moral *see* moral identity approach
   political 96
   shared sense of 98
immigration 94–8
   defending limits on 97–8
   and distributive justice 95–8
   open 95, 97
   policy 96–8
   protecting fellow citizens from
      negative effects 95–6
impartiality 71–3
   strict 97–8
   and utilitarianism 74, 76–7
imperfection, politics of 3, 109–19
inequalities 98–9, 103–4, 116–17,
      130–1
   economic 100–1, 105–7, 109,
      113–14, 140
   genuine choice excusing "otherwise
      unacceptable inequalities" 100,
      102, 110
   luck egalitarianism 100, 111
   social 100–1
intrinsic values
   see also accepting-appreciating stance
      (accepting the given); values
   good life, living 133–4, 137
   responsiveness to 133–4, 142–3
   universal mastery violating

   accepting-appreciating stance
      17–20, 26–7
   contradicting spiritual
      requirements 17, 20–1
   and virtues 1–2
irreplaceability 41–2

Jefferson, Thomas 90–1, 145–6, 155
Jesus Christ 74, 104
   Gospel of Matthew 138–9
   Parable of the Good Samaritan 74
   "Sermon on the Mount" 138–9
John Paul II, Pope 138, 150
   Laborem Exercens 148–50
Judaism 28–9, 73–4
   Sabbath-orientation/practice 156–8
   Talmud 141
justice
   see also distributive justice;
      sufficientarian justice
   Cohen on 115–16, 118
   as desert (traditional common-sense
      notion) 102, 107
   ideal theory 111
   limiting virtues constraining
      conception of 118–19
   need for 114
   nonideal theory 111
   "Platonic" versus "Aristotelian"
      conception of 115–16
   seen by Hume as a jealous
      virtue 114–15

Kahane, Guy 18–20, 42
Kantianism 3, 71–2, 103
Karamazov, Ivan 9, 21
Kass, Leon 18–20, 46–9, 60–2, 79
   The Hungry Soul 47, 50, 59–60, 132,
      134–5
Kekes, John 105–7
Keynes, John Maynard 129–32,
      135, 148
   "Economic Possibilities for Our
      Grandchildren" 129
Kleinig, John 70

koinōnia 98, 114–15
Kolnai, Aurel 24–5, 30, 116
Kukathas, Chandran 94–5, 97–8

Lasch, Christopher 89, 125
    The True and Only Heaven 124
Leiss, William 6–7
leisure 129–32, 157–8, 160–1
Leo XIII, Pope
    Rerum Novarum 148–50
li (ritual, custom, ceremony, good
        manners) *see* Confucius/
        Confucian way of life; good
        manners
liberalism 124–5, 132–3
liberation, sense of 9
limited government 123
limiting virtues 1–4, 15–16, 18–20, 118
    see also contentment; humility;
        loyalty; moderation;
        neighborliness; reverence/
        respect; virtue(s)
    defining 1–2
    and limits *see* economic limits;
        existential limits; moral limits;
        political limits
loneliness 77–8
love 8, 10–11, 41–2, 89, 94
    capacity for 88
    in families 18–20, 80–1
    of God 70
    for humanity 75–6, 86–7, 91–2
    of money 129–32
    of neighbor 8, 73–4, 77–8, 86–7
    romantic 56, 62, 78–80
    self-love 128
loyalty
    see also neighborliness
    in family relationships 79–81
    to fellow citizens 96
    to friends 82–3
    to the given 40–5, 77
    to humanity 77
    limits of 81–2
    to nation-states 92–4

to one's country *see* patriotism
patriotic 112–13
luck egalitarianism 42, 100, 105–7
    defining 100
    as account of distributive justice 98–102
    and existential conservatism 107
    and the given 103–4
    and sufficientarianism 98–100, 103–4
    and utopianism 111, 116

McCarthy, Cormac
    No Country for Old Men 51, 53–4
McClure, Jane Michele 154
Madison, James 121–3
magnanimity 28–9
Mandeville, Bernard 127–8, 135
    The Fable of the Bees 127
Manichaeism 120–1
manners
    see also good manners
    bad manners, as opposed to good
        51–2, 54
    table manners 60–2
market reciprocity 112
Marx, Karl 130–1, 151–2
    The Communist Manifesto 145–6
May, Simon 26–7
Meilaender, Gilbert 18–20, 80–1
Mencius 51–2, 71–2
meritocratic hubris 142–3
    see also hubris (improper pride)
Merry Wives of Windsor
    (Shakespeare) 18, 140
Merton, Thomas 154
Mill, John Stuart 71–2, 74
Miller, David 96–8, 110–11, 117–18
minimalism 134–5
Mitchell, Mark T. 88–90
moderation
    character formation 2–3, 48, 57–62
    and limits of government 119–25
    as a master virtue of character 2–3
    and temperance 57–8
    as a master virtue of character 2–3,
        57–8

money
  and happiness 136–7
  love of 129–32
  sufficiency of 136–7
  and wealth 126–7
moral egalitarianism 99
morality
  see also moral limits
  of compassion 10–11
  moral identity approach 65–7
  "morality itself" 65–6
  moral repugnance 18
  preventing moral anarchy approach,
    absolute prohibitions 63–7
  utilitarian 74
moral limits 46–83
  see also absolute prohibitions
  character formation 46–62
  duties of assistance 71–83
  and humility 33
moderation 57–62
multiculturalism 98
murder 65–7
  see also dignity of human life
  prohibition on 69–70
Murdoch, Iris 158–9

Nagel, Thomas 104–7
nationality principle 95
nation-states 92, 94, 97–8
  see also distributive justice; patriotism
  and distributive justice 97
  loyalty to 92–4
  welfare state 95
nature/natural world 5–7
neighborliness 46, 71–83
  see also Parable of the Good
    Samaritan
  concrete humanity, solidarity with 75
  as important limiting virtue 73–4
  love of neighbor 8, 73–4, 77–8, 86–7
New Testament 20, 137–9
Nietzsche, Friedrich 7–12
  accepting-appreciating stance,
    appearing to adopt 25–7

  on contentment 34–5
  Dworkin compared 14–15
  on ethic of power 10–11, 25–6
  on growth in power 10–12, 21, 26–7
  May's evaluation of 26–7
  on playing God 9–10
  and primacy of choosing-controlling
    stance 14
  and Promethean ideal 7
  on "self-overcoming" 22–3
  on suffering 21
  threat of nihilism 8–10, 25
  on will to power 10–11, 21, 26–7
  works of see *The Gay Science*,
    Nietzsche ("death of God"
    passage); *Thus Spoke
    Zarathustra* (Nietzsche)
nihilism
  see also Dworkin, Ronald; Nietzsche,
    Friedrich
  and choosing-controlling stance 23
  courting of 23–5
  and despair 8–10
  moral 57
  practice of 10–11
  of Schopenhauer 10–11
  subjective and objective 23
  threat of 14, 26–7, 32, 42
    for Dworkin 25, 27–8
    for Nietzsche 8–10, 25
non-existence 37–8
Nozick, Robert 88, 102–3, 110, 135–6
Nussbaum, Martha 28–9, 43, 75–6,
    85–7, 89, 94
  "Patriotism and
    Cosmopolitanism" 85, 88
  "Transcending Humanity" 88

O'Connor, Flannery 57
oikonomia (household
    management) 126, 143–6
open borders 94, 98

Parable of the Good Samaritan 74
Parens, Erik 15

parents
  bad 54–5
  filial piety 30–1
  gift of life given by *see* gift, human life
      perceived as
  good 55–6
  loyalty owed by children to 80–1
  loyalty to own children 79
  parental love 18–20
  parent–child relationship 18–20
  parent-surrogates 54–5
  respect owed to 30–1, 51–2, 54–6
patriotism 77, 84–7
  false forms 93
  and love of place 87–8
  loyalty to nation-state 92–3, 112–13
  and need to belong 88
perfection and imperfection 21–2
Pieper, Josef 60–2, 160
piety
  see also filial piety; reverence
  humility and reverence 32
  and impiety 65–6
  religious 30–1
Plato 79, 115–16
  Euthyphro 81–2
Platts, Mark 24–5
playing God
  see also choosing-controlling stance
      (seeking improvement); God;
      Promethean ideal (seeking
      mastery over the given);
      Prometheanism
  Dworkin on 12–14
  and genetic engineering 12–13, 21
  history 6–7
  Nietzsche on 8–11
  as playing with fire 14
  Promethean ideal 2, 5–7, 32
  as solution to "death of God" 8–9
pleonexia (overreaching in benefits for
      ourselves) 114–15, 131–2
polis (political community) *see*
      citizenship and political
      community

political community *see* citizenship and
      political community
political decentralization 89–90, 92–3,
      98, 145–6
political limits 84–125
  community 84–98
  cosmopolitanism *see* cosmopolitanism
  imperfection 109–19
  sufficientarian justice 84, 98–110
  universal versus particular 86–8
positive freedom 135–40
power
  see also Nietzsche, Friedrich
  absolute 159–60
  distribution of 121–2
  economic 75–6, 90–1
  effective 135–8, 140
  ethic of (Nietzsche) 10–11, 25–6
  governmental 90–1
  growth in (Nietzsche) 10–12, 25–6, 34–5
  of large corporations 154–5
  neutral 121–2
  political 75–6, 92
  to prevent harm 72–3
  transformative 5–6
  will to power (Nietzsche) 10–11, 21,
      26–7
preferences 137, 143
  preference maximizers 143
  second-order 135–6
preference-satisfaction 64
preventing moral anarchy approach,
      absolute prohibitions 63–6
  consequentialism 64–6
  utilitarianism 64
pride 28–30, 33–4
procedural republic 90–1
progress, belief in 124–5
progressive optimism 124–5
Promethean ideal (seeking mastery
      over the given) 5–15
  see also choosing-controlling stance
      (seeking improvement);
      Dworkin, Ronald; Nietzsche,
      Friedrich

Promethean ideal (seeking mastery over the given) (*cont.*)
    and Dworkin 14–15
    and genetic engineering *see* genetic engineering of children
    and modernity 6–7
    and Nietzsche 7
    opposition to 15–28
    playing God 2, 5–7, 32
    science and technology 5–7
    versions of 6–7, 21
Prometheanism
    see also Promethean ideal (seeking mastery over the given)
    consequentialist line of argument against 17
    genetic engineering 12–13
    myth of Prometheus 5–6
    and playing God 14
    positive freedom 140
    pursuit of achievement 22–3
    scientific-technological version 7
    and wealth 140–1
proximity 73, 118–19
    see also neighborliness
    moral significance 3, 46, 73–4, 86–7
Pugh, Jonathan 42–3

Quinton, Anthony 109–10

radical contingency problem 66–7
rational choice model of human behavior 143
rationality 47–8
Rawls, John 90–1, 99–105, 132–3
relational egalitarianism 99–101
religion
    see also Christianity; Genesis, Hebrew Bible; God; Judaism; New Testament; Ten Commandments
    secular 124
    true 21–2, 36, 161
republicanism 90–1

respect *see* reverence/respect
restraints on desires 1–2, 46–9, 137
    see also desires
    and reverence 62–71
reverence/respect 51–7
    see also dignity of human life; humility
    and absolute prohibitions *see* absolute prohibitions
    character formation 48, 51–7
    versus dominion stance 30–1
    and good manners 52, 56
    and humility 2, 28–34
    and irreverence 33–4
    and justice 118–19
    owed to parents or elders 30–1, 51–2, 54–6
    respect for life 69–70
    respect-worthiness 53, 56–7
    responsiveness 32
    and restraints on desires 62–71
    reverence as a heightened form of respect 30–1
    reverence-worthiness 30–3, 51–3, 57, 63, 65–6, 69–71
    Woodruff on 32–3
rituals
    character formation 52–3, 56–7
    religious 48–9, 56
Roberts, Robert C. 29–30, 60–2
Robinson Crusoe **(Defoe)** 37–8
romantic love 56
    versus lust 62, 78–9
rule-utilitarianism 68

Sabbath-orientation/practice 22–3, 148–50, 155–61
    double meaning 159–60
sacredness
    see also reverence/respect
    good manners 48–9
    reverence/respect 54–6, 62, 70–1,
        and absolute prohibitions 63, 66–7, 69–70
        and humility 30–1

Sandel, Michael 25–6, 29–31, 92–3, 99, 103, 142–3, 159–60
  The Case against Perfection 15–16, 18–20, 24, 90–1
  Democracy's Discontent 90–1
  What Money Can't Buy 130–1
Savulescu, Julian 12–13, 42
Scheler, Max 119
Schopenhauer, Arthur 10–11, 21
science and technology 5–7, 12
Scruton, Roger 78–9, 108–9, 116–17, 143–6
self-creation, projects of 9
self-government, democratic 84, 89–90, 92, 94, 98
  and cultural self-determination 96
self-love 128
separation of powers 123
shame 58–9
Shelley, Mary
  Frankenstein 17
Sher, George 103–4
Shields, Philip R. 44
Shiffman, Mark 144–5
Singer, Peter 72–3, 75–7
  "Famine, Affluence, and Morality" 72–3
Skidelsky, Robert and Edward 126–7, 130–7
Smith, Adam 135
  invisible hand 127–9
  The Theory of Moral Sentiments 127–8
  The Wealth of Nations 128
socialism
  versus capitalism 135–6
  Cohen on socialist ideal 16–17, 111–16, 129
  feasibility problem and socialist ideal 113–16
  utopia 114
social status 130–2
solidarity
  distributive justice 97, 101–2
  with fellow citizens 87

neighborliness and proximity 3, 46, 73–5, 86–7
patriotism 96, 112–13
social 142–3
sufficientarian justice 104
sovereignty, dispersed 89–90, 92–3
spiritual requirements, stance of universal mastery contradicting 17, 20–1
steady-state economy 135–6
Stoicism 85–6, 88
subsidiarity principle 89–90, 98
suffering 10–11, 20–1
sufficientarian justice 3, 84, 98–110
  and accepting-appreciating stance 108
  and egalitarianism 98–9, 107–8
  sufficientarian conception of distributive justice 98–9, 108–9, 112
suicide 44–5

Talmud 141
Taylor, Charles 6–7
temperance
  and intemperance 59–60
  and moderation 57–8
  as "utilitarian" virtue 69–70
Ten Commandments 156–7
theism 18–20, 30–1, 40, 69
Theogony (Hesiod) 5–6
theological voluntarism 69
Thus Spoke Zarathustra (Nietzsche) 7, 26–7
  "last man" 11–12
  "overman" 9–12, 34–5
Tocqueville, Alexis de 90–1, 99
Tower of Babel (Genesis 11:1–9) 6–7
transhumanism 10, 12, 27–8, 34–5
tribalism, antagonistic 88

United States
  Constitution 89–90, 121–3
  displacement, in American history 146–7

186 INDEX

United States (*cont.*)
  federalism 89–91
  patriotic loyalty to 92–3
  republic 90–1
  Tenth Amendment 89–90, 123
usury 129–30
utilitarianism
  absolute prohibitions 64
  duties of assistance 71–3
  and impartiality 74, 76–7
  and love of neighbor 74
  and Rawls 103
  rule-utilitarianism 68
  and temperance 69–70
utopianism 84, 116–17, 119, 124–5, 129
  and capitalism 125
  critiques 116–19
  fundamental complaint against 116
  and luck egalitarianism 111, 116
  and socialism 114
  utopian ideal 124–5
  utopian schemes 117–19, 123, 145–6
  utopian visions 124–5, 154–5

values
  and attachment 43
  autonomy principle 27–8
  beneficence principle 27–8
  choosing 90–1
  conflicting 104–5, 118
  creativity 9
  defense of existing value 41
  intrinsic *see* intrinsic values
  object of value 41–2
  particular valuing 16–17, 41–2
  personal valuing 16–17, 43
  public culture 96
  re-evaluating 8–9, 14–15
  and Sabbath-orientation 158–9
vices 29–30, 59–60, 67–8, 119, 137–8
  see also greed, vice of; hubris
    (improper pride); limiting
    virtues; virtue(s)
  discontentment 35
  envy 105–7

fear, excessive or deficient 57–8
giving, excessive or deficient 57–8
intemperance 59–60
pleonexia (overreaching in benefits for
  ourselves) 114–15, 131–2
virtue(s)
  see also limiting virtues
  of character, cultivating 127
  charity 67
  and Confucian ethics 48–9
  expressing 48–9, 51–2
  frugality 134–5
  genuine 38, 81–5
  and good manners 50
  necessary for living a good life 1–2,
    133–4, 144–5
  sacrificing for economic gain 130–1
  self-sacrifice 144
  thrift 144
  traditional virtue ethics 135, 137–8
  unity of 81–2

Walzer, Michael 95–8
wealth
  see also social status
  capitalism and desire for wealth 130–1
  desire for 130–1
  extreme riches 140
  and freedom 135–6, 140
  insatiable desire for 130–3, 136–7,
    141–2
  and money 126–7
  putting to civilized use 130–2
Weckert, John 17
Weil, Simone 88
Wiggins, David 23, 25–6, 36–7, 103–4,
  107, 137, 148–50
  on "metaphysical crusade against
    contingency" 99–100, 104–8
  "Neo-Aristotelian Reflections on
    Justice" 114–15
wilfullness 141–2
will
  constraints/limits on will 18, 24,
    26–7, 30–4

having nothing to will 24–5
will to power (Nietzsche) 10–11, 21, 26–7
willfulness 15–16, 18, 141–2
Williams, Bernard 17, 66–7, 71–2
Wirzba, Norman 158–9
Woodruff, Paul 32–3
work, importance of 147
    see also leisure
    alienated labor 151–2
    dignity of work 126, 148–50

Works and Days (Hesiod) 5–6
world citizenship 86, 94
wrongful acts, negative duties of
    avoiding 71

Yeats, W. B.
    "A Prayer for My Daughter"
        50, 62
    "Sailing to Byzantium" 51
    "The Second Coming" 50–1